ROBERT FULTON

ROBERT FULTON

BY JOHN S. ~~S.~~ ^mith MORGAN

MASON/CHARTER   NEW YORK 1977

Library of Congress Cataloging in Publication Data

Morgan, John Smith, 1921–
Robert Fulton.

   Bibliography: p.
   Includes index.
   1. Fulton, Robert, 1765–1815. 2. Marine
engineers—United States—Biography. 3. Inventors
—United States—Biography.
VM140.F9M67   623.82'4'0924 [B]   77-633
ISBN 0-88405-438-1

CONTENTS

ACKNOWLEDGMENTS

Of the two men who had the greatest influence on Fulton—Joel Barlow and Chancellor Robert R. Livingston—my interpretation of the former's role in the engineer's life is based largely on material from and interviews with the great, great grandnephew of the first Joel Barlow. He is the second Joel Barlow, an attorney living in Washington, D.C.

By fortunate coincidence, Mrs. Joel Barlow, the wife of the second Joel Barlow, is a descendant of the chancellor's family. She and her mother, Mrs. Philip Livingston Poe, provided insights about his experiences with the steamboat. Most biographical material already recorded about the chancellor concentrates on his political and diplomatic roles. I attempt to make clear his important part in making the vessel a commercial success.

Four libraries supplied the bulk of the source material: the Montagu Collection of the New York Public Library, the Fulton Collection of the New-York Historical Society, the Library of Congress and the Carnegie Library of Pittsburgh.

I am grateful to all for their help and especially for that of my wife, Virginia W. Morgan.

JOHN S. MORGAN

PROLOGUE

When Robert Fulton died in 1815, he probably never would have mentioned his fundamental accomplishments:

Inventor of the modern concept of invention. He went at it systematically and methodically, as a pursuit in itself—a notion foreign to his contemporaries and predecessors. They regarded an invention as a showpiece or abstraction, to be displayed like a circus act or described in journals as "philosophy." The invention's use was a secondary consideration. Above all else, Fulton had a pragmatic approach to invention. He wrote about his developments, but only to publicize them and to earn money from them. For him, invention was a business by which he earned his living. If a canal-digging machine wouldn't sell, he turned to a rope-making device that would find a market. He had finely honed his sense of what the world needed.

Advocate of research and development. Most people up until the early nineteenth century saw the world as static. Few accepted the idea of deliberately spending time, money and energy to improve something that already existed. He exasperated his partners by "extravagant" spending on devices that they thought were already good enough. Fulton's contemporary, James Watt, who surpassed him in engineering skill, retired early after inventing the steam engine and discouraged its application in the new-fangled steamboat. Fulton kept everlastingly at improving his steamboat and other inventions. He maintained records and ran tests, unheard of in his time. Conventional wisdom had it that inventions sprang in full bloom from the inventor's head. Fulton knew that they took work, time and money. He never stopped spending his own or others' treasure. To this day, the debate persists on the merits of spending for research and development.

One of the world's first technologists. Technology consists of the sum of the ways in which a society provides the material things for its civilization. Fulton gathered devices and objects together and made something new of them, at the same time showing the world how to use it. Everything about the steamboat had already been invented—the boat, steam engine, paddle wheel, connecting devices. He brought them all together into a mechanically propelled boat. At the same time, he did something else essential for a technologist: He sold the public on his invention's utility and safety, an achievement in which all his steamboat predecessors had failed. Indeed, some of Fulton's engineering contemporaries sneered at this quality—"a mere peddler," they jeered. Yet, what good is a steamboat if no one will use it? As part of his approach to technology, Fulton looked at markets for his invention. Although he made mistakes in his market research (as still sometimes happens today), he recognized early the importance of predicting who the users for his invention would be. As another aspect of his role as technologist, Fulton began an approach to systematic manufacturing in America. Until then, one part for a final product was made here, another there. Fulton tried to produce as much of the finished product as possible under one roof, training his own craftsmen in the work. Such organization was unique in 1815.

One of the world's first military technologists. Although we may not thank him for it, he helped introduce technological warfare into three countries—France, Britain and America. By trying to sell the submarine and torpedo to three different military establishments, he did much toward awakening the nations to the uses of technology in modern warfare. He began the shift to steam warships and urged that America build ironclads nearly a half-century before the first one steamed into battle.

One of America's first engineers. As a technologist, Fulton had to be an engineer. The profession was virtually unknown in America and rare in England where its origins lie. Although without formal education as an engineer, he taught himself the profession and helped teach America its importance.

Tamer of the wilderness by machines. He epitomized a new kind of pioneer. He accurately foresaw how improved transportation would do more to open the West than a thousand Daniel Boones. He built the first steamboat that chugged down the Ohio and Mississippi rivers.

Why did Fulton underplay these contributions, although his

enemies tagged him as a braggart and even as a charlatan who claimed many accomplishments that were not truly his?

One explanation is that "technology" and "technologist" had not even been coined in 1815. Fulton never called himself a technologist because the word didn't exist, nor was the concept known. He sometimes named himself engineer, but purists objected because he lacked a formal education in engineering—or much else.

Another explanation lies in snobbery. The Pennsylvania farm boy, son of a tailor, turned himself into a gentleman and was proud of the achievement. In 1815, American gentlemen didn't build things. They practiced professions, managed land, or engaged in government or shipping. He called himself a shipowner who built them only because no one else could meet his exacting specifications.

Robert Fulton wanted the world to know that he was a gentleman who built a fortune out of a shipping line. As it happened, neither his fortune nor his shipping line long survived him. But he did succeed in starting an engineering and technological revolution that helped change the face of this earth and continues to do so to this day.

<table>
<tr><td>**1**</td><td>## BEGINNING IN LANCASTER</td></tr>
</table>

Robert Fulton, Jr. was born on November 14, 1765 in Little Britain Township, Pennsylvania. It was a time of uncertainty for his parents on their recently purchased farm. His father, a moderately successful tailor, had decided to abandon his trade and become a farmer, a decision that already must have looked mistaken after his first poor harvest.[1]

Robert Fulton, Sr. was a Scotch-Irishman who had emigrated from Kilkenny County in Ireland to Philadelphia.[2] Definite word first locates him in Lancaster about 1735, as already a tailor and with a little trading business as a sideline.[3] He would have had to develop a sideline because Lancaster had a population of only two hundred in 1750. (Even by 1765, when its population had tripled to 600, that number of people would scarcely have kept a tailor fully occupied even if he had all their business.)

Fulton was a solid citizen—one of the founders of the First Presbyterian Church in Lancaster, secretary of the Union Fire Company, assistant burgess of the borough, and a charter member of the Juliana Library of Lancaster which was established in 1763 as the third library in the American colonies.

In 1759, he had married Mary Smith,[4] the daughter of an impoverished Scotch-Irish Presbyterian farmer.

For the first years of his marriage, Robert Sr. stuck to his needle and thread. He was doing well enough to bring his bride home to a brick house he had purchased on August 23, 1759 on the northeast corner of what was then called Centre Square, now known as Penn Square.[5] During their stay of five and one-half years there, his wife bore three daughters—Elizabeth, Isabella (called Bell),

1

and Mary, nicknamed Polly to distinguish her from her mother.

Yet the tailor either didn't have enough business, grew restless, or both. At a foreclosure sale on February 8, 1765 he bought a farm of nearly 400 acres [6] raising the large sum of £965 (about $40,000 in today's purchasing power) by selling his house and shop and by borrowing from three Philadelphians, one of them named Joseph Swift. Later he must have borrowed still more from them because court records show that his indebtedness had risen to £1777.9.10 (about $75,000) by November 29, 1766.

The property lies 30 miles south of Lancaster near the Maryland state line. The farmhouse still stands, built of brown and gray stones held together with lime mortar. Facing east, overlooking pleasant rolling Pennsylvania hills, it has three stories, three rooms on the first floor, two on the second, and one on the third, with generous storage space. The Fultons moved there in April, the traditional moving date in colonial America—especially for farmers who had to begin their spring planting.

The first son, Robert, Jr., was born in the room above the parlor. His birth was probably the only success the Fultons enjoyed in 1765. Nor did much success follow for the next six years while the family struggled on the barren acres. Young Fulton, a precocious boy, may have absorbed memories of defeat and despair. He later told friends that he "attached no importance to the circumstances of my birth." [7] In his letters and other writings, references rarely surface concerning his ancestry, birthplace or Lancaster.

In 1771, the Fulton name no longer appeared on the Little Britain Township tax lists; the tailor-farmer couldn't meet his mortgage payments and Joseph Swift took the place over. Swift and his descendents lived there until 1966. (The three surviving Swift sisters then sold the home to the Commonwealth of Pennsylvania, which restored it and maintains it as an historical site.) In 1844, a section, including this property, was cut off from Little Britain Township and named Fulton Township in honor of the son who was born and lived the first six years of his life there. Might he not have found the nature of the recognition ironic?

The Fultons returned to Lancaster where the father tried to reestablish himself as a tailor. Three years later—before he had time to succeed—he died, without making a will, no doubt because he had little or nothing to leave.

The Widow Mary Fulton had five children to support because a fifth child and second son, Abraham, had been born. She sent some of the children to live with her relatives. Nine-year-old Bob went to his Uncle Joseph Smith for a short time, but he was unhappy and returned home. Joseph Smith had become a Presbyterian minister, and the boy, already nick-named "Quicksilver Bob" [8] because of his mercurial enthusiasms and his experiments with mercury, may have rebelled at the somber austerity of his uncle's household.

His mother had a more permissive temperament. A lively, voluble woman, she passed on those traits to her son. Well educated for her time, she taught him herself at home until he was eight. Nor did she require him to follow all the rigid religious observances so common in that era. However, he did continue to attend services in the church founded by his father—whose stentorian hymnal renditions were still remembered.

Perhaps the mother found permissiveness the wisest course with Robert who was proving something of a rebel. She sent him to a school taught in Lancaster by Caleb Johnson, a Quaker, a Tory, and a man of narrow and traditional views on teaching as well as politics. Johnson and Fulton didn't hit it off. As was the custom, Johnson didn't spare the rod. After one caning session on the boy's hands, the youngster cried, "Sir, I came to have something beaten into my brains, and not into my knuckles." [9]

On another occasion when young Fulton arrived late to school, he escaped punishment by showing what had caused his tardiness—he had made lead for pencils. The teacher was so impressed by the excellence of the product that he forgot to cane him and forgave him.

In the Caleb Johnson school, a classmate had brought colors mixed in cockle-shells. Fulton, who could wheedle, if not beg outright when he wanted something, got half. He soon showed that he could paint so much better than his friend that he eventually got all the colors. He began drawing and painting with an intensity that later marked all his enthusiasms. For a while, he forgot everything else, including school.

When Mrs. Fulton had a conference about her son with Johnson, she probably didn't know whether to be glad or sad when the teacher reported that the boy had told him: "My head is so full of original notions that there is no vacant chamber to store away the contents of dusty books." [10]

Fulton did get the rudiments of education, but the syntax and spelling in his early letters and writings that survive are loose even

by the casual standards of the eighteenth century. When he was
twenty-four he would still be writing his mother, from London,
like a schoolboy: "In your next letter please to give me a very per-
ticular account of everything you know perticularly how you like
the little farm if you have a good garden and what kind of Neigh-
bours you have got." [11]

Fulton's real education came in the streets, alleys and work-
shops of Lancaster. As in most of colonial America during the
Revolution, formal education practically ceased in 1775. Johnson
found himself in trouble because of his Tory beliefs and probably
would not have been allowed to teach even if any parents had
agreed to send him students.

Lancaster became famous as a depot of supplies for the American forces during the Revolution. Rifles, blankets, clothing and
gunpowder were manufactured there. At the time of the Revolution, it was the largest inland city in the thirteen colonies, its population swelled by refugees from coastal areas. Philadelphia, then
the largest metropolis in the colonies and the major intellectual
center, was seized by the British, and sent its great to the inland
town sixty-five miles to the west. On September 27, 1777 the Continental Congress sat in Lancaster for the one day. Young Fulton
may have seen the likes of John Adams, John Hancock, Thomas
Paine and David Rittenhouse.

And Lancaster had some of its own great: William Henry, for
example. Born in Chester County in 1729, he soon came to Lancaster where he began making the Kentucky rifle, invented in that
town. He performed as an armorer during the French and Indian
Wars. By the time of the Revolution, he was the major gunmaker
in America, so important that a sentry guarded his door. It's unlikely that a poor young boy ever got near him.

Yet it is tempting to speculate that he did because Henry definitely played with the idea of steamboats, mentioning the possibility to a mathematician, Andrew Ellicott, in 1775 and to Thomas
Paine in 1778. In the 1780's, he finally found time to put his ideas
on paper, but he evidently went no further.

He had many other more promising projects—labor-saving machinery for his gun works, the discovery of the screw auger, and
the invention of the "sentinel register" (a device using heated air
to automatically open and close furnace flues). Besides this, he
was treasurer for Pennsylvania, and he performed many other
duties.

This western town seethed with invention and mechanical expertise. To make the Kentucky rifle, a necessity for the pioneers heading for the wilderness, required sophisticated technology. This developed in Lancaster, which at this time boasted some of the best mechanics in the colonies.

If "Quicksilver Bob" had no contact with William Henry, he may have known lesser known gunsmiths, such as Isch or Messersmith. An ingratiating boy could make himself pleasant and useful to the harried mechanics working day and night and even on Sunday, a rarity in those days. If an eight-year-old boy could master the art of making lead for pencils from Nicholas Miller, a town craftsman, he could pick up other rudimentary skills to help out in gunmaking two years later.

By 1779, Fulton knew enough to design a novel air gun, with sufficient promise for manufacture in a pilot model. Although it must have failed as an invention—because nothing more is heard of it, the project indicates his growing technical expertise.

Fulton had also developed his other useful skill—drawing. He picked up a few pennies painting signs for tradesmen. At that time the streets of Lancaster were like picture galleries, dotted with numerous painted signs as the shopkeepers tried to outdo each other with colorful advertisements.

The practice had already been established for some years. An earlier artist, Benjamin West, had supplemented his income as a sign painter when he had lived for a while in Lancaster.[12] But at this time, Fulton had not met West, who was twenty-seven years older than he and already pursuing a successful artistic career in Europe. West had been born in Chester; under eighteenth century traveling conditions a two days' ride from Lancaster.

The West-Fulton connection was yet to come. The older man became one in a series of Robert Fulton's patrons, and West's nephew, David Morris, was to marry Robert's sister, Polly. Yet West had made his mark in Lancaster, and it's likely that his example encouraged Fulton to follow a career in art.

Another romantic story tells us that Major John André, the British spy, taught Fulton and other Lancaster boys how to paint. Possibly it's true, although Fulton never mentions it in any of his writings. It is known that André was kept a prisoner in Lancaster for a time. In those more casual days, imprisoned officers were allowed the freedom of the town, provided they promised not to leave. André lodged with the family of Caleb Cope who had five sons, all of whom Fulton knew, and André did dabble as a painter.

Whether Fulton taught himself or had help from others, eleven-year-old Robert was skilled enough to make mechanical drawings for the gunsmiths. He also suggested designs for the stocks on the Kentucky rifles (for some were still decorated—war or not).

The shops and streets of Lancaster taught Fulton another side of war. The Americans used this inland city as a camp for prisoners. He was ten years old when he saw his first British sailors marching in ragged lines under American guards. Officers lived almost as free men, but enlisted men existed in the local barracks or in makeshift shelters. Eventually, many Hessians were imprisoned in Lancaster. Bewildered peasants who spoke no English had been pressed into service and didn't know what they were doing so far from home. Robert drew a caricature of them, showing them being annihilated by a rope. This conception arose from the fact that the town officials had erected a rope beyond which the prisoners could not come at night's curfew time. The drawing was cruel: the Hessians and their band offered much entertainment to the townfolks, especially to the boys. On June 4, 1777 prisoners in the barracks disarmed their guards. Church bells sounded the alarm and militia men rushed in and saved the situation.

Messengers came frequently with sensational news—that Philadelphia had fallen to the British, that it had been retaken by Washington, that Howe had been driven back to his ships. Or if news about the war against the British was dull, rumors surfaced regularly about possible Indian raids. Many a day or night rang with the clamor of militia assembling in the town square to meet the threats that usually proved false. But what a time for a young boy to be growing up! As an adult, Fulton was almost constantly on the move, forever embarking on a new adventure. Could some of this characteristic been bred into him during his youth when he lived for years on a nearly perpetual diet of excitement and urgency?

Years later, in a rare letter of retrospection, he recalled to his sister, Elizabeth:

> I remember when we all lived in Lancaster . . . I was a stripling of
> a boy about 12 years old and our mother, being at Mr. Craig's in

the country, I had a battle against the two sisters. You and Bell had turned me and the cat out of the truckle bed. It was a winter evening about 8 o'clock. I instantly flew to the tongs and as I stood in my shirt with uplifted arms ready to knock all your brains out you were so much astonished at my resolute manner and wickedness, as you supposed, that you began to cry & said you were sure I should some day be hanged.[13]

The girls' tears melted his anger, but the incident illustrates the temper of the boy—and of the times.

One wonders if he could have indulged in any quieter boyish pursuits. No evidence survives that he did, so some biographers may have invented them. For example, the tale persists that he went fishing with an older boy named Christopher Gumpf, an apprentice at Isch & Messersmith; that the two grew tired of rowing the heavy boat, so Robert invented a stern paddle wheel arrangement that the boys pumped with their feet. Allegedly, it worked. What makes the story suspect lies in the fact that Fulton never mentions it, not even when, in a court fight years later, he was trying to establish the validity of his patents. Surely he would have dredged up this youthful experiment to help him establish priority, if he could legitimately have done so. He did cite other tenuous "proofs," such as a decades-old letter whose validity turned out to be doubtful.

Another story relates that young Fulton used gunpowder to replace the scarce candles (surely as scarce as candles!). The only basis for this yarn is the fact that during celebrations of American victories every patriot had to show light in his windows or be branded as a Tory and have them broken by the drunken militiamen. In an emergency, a desperate householder may have had to use for light whatever lay at hand. A more likely story has young Fulton concocting primitive rockets made from gunpowder and pasteboard to celebrate the July 4 signing of the Declaration of Independence.

War, a chance at art, mechanical opportunities, searing poverty—those were the factors that educated Robert Fulton in Lancaster. He would spend his adult life largely in inventing devices for war. Until he was in his late twenties, he tried to be an artist. His mechanical genius united with his art to produce inventions. The fear of poverty drove him all his life; from the day he discovered he was poor, he wanted to be rich.

2 APPRENTICESHIP IN PHILADELPHIA

Sometime between 1779 and 1782—but probably closer to the earlier date—Robert Fulton left Lancaster for Philadelphia. He moved only 65 miles geographically, but he entered a different world—one rich in arts and crafts, society, and intellectual stimulation.

His mother had apprenticed him to a jeweler, Jeremiah Andrews, a Londoner who had migrated to New York in 1774 and moved on to Philadelphia.[1] According to his Philadelphia advertisements, Andrews carried:

> a neat assortment of jewelry of his own manufacture equal to any imported. Ladies' paste shoe buckles of the newest patterns, gentlemen's knee and stock buckles, garnet hoop rings and every other kind of rings, lockets and brooches, plain or with stone, hair worked in the neatest manner, mourning rings made on the shortest notice, with every other kind of jewelry by their humble servant, Jeremiah Andrews, living in Second Street between Chestnut and Walnut Streets, Philadelphia.[2]

Mrs. Fulton or her advisors had chosen wisely for young Robert. The jeweler was the élite craftsman of the late eighteenth century, a recent offshoot from the silversmith. In one craft he combined both the skills of the artisan and the artist. Young Fulton showed promise in both disciplines. Furthermore, Andrews evidently proved to be a good master, both in teaching Fulton and in his treatment of him.

Biographies of the colonists are strewn with horror stories of bad experiences as apprentices. Ben Franklin couldn't endure his

8

own brother as his master and ran away from a printing apprenticeship. John Fitch, whom Fulton would soon encounter, was psychologically scarred for life by his miserable apprenticeship.

Although the system was breaking down by the time young Fulton embarked upon it, it still remained one of the principal organized methods of instruction for youth, supposedly involving supervision of religious and other studies as well as the teaching of a trade. But because apprenticeship was also used as a penal method for debtors, idlers, the unemployed, and as a method of poor relief, the youthful apprentice often found himself classed with delinquents.

If a master went bankrupt, he listed his apprentices among his assets. If the master died, the apprentice's remaining period of indenture was often sold with the rest of the estate. Yet, Mary Fulton on September 16, 1774 renounced before the Lancaster Register's Office the only tangible asset of her husband's estate, Samuel Chapman, an eleven-year-old pauper who had been apprenticed to Fulton Sr. in 1772 to learn tailoring. The overseers of the poor were then asked to assume responsibility for Chapman because the estate could not sell his indenture and did not have the funds to continue his apprenticeship.

In America's colonial era and shortly thereafter, the apprenticeship period commonly ran for seven years, usually age fourteen to twenty-one. However, many exceptions occurred, especially during the Revolutionary War. In England, which served as the model for American practices, the legal requirements which made an apprentice a virtual indentured slave, were repealed in 1814. In America, apprenticeship along the British model had largely decayed by the same date. But in Fulton's time the system still survived. Fortunately, he fared well under it.

He learned much from his new surroundings, too. All his life, he learned more from what went on around him than from books. This was partly because he remained poorly educated until his early twenties, partly because books and newspapers were critically scarce during this time in America's history, and partly because he was naturally a visual and oral person. As an embryonic artist and engineer, he learned best by seeing, doing, talking and listening. Already a voluble talker, he had the rarer gift of listening, at least when something interested him—and much did in Philadelphia.

Although the city had been occupied by the British for nine months from September, 1777 until June, 1778, it had never suffered the physical destruction inflicted upon New York. It quickly resumed its position as the leading city in America, with a population of nearly 43,000.

Close-flanking red brick houses faced each other across narrow streets. Although the Quakers had driven the theater beyond the city limits, they had generously welcomed all manner of learning. By its geographical position on two navigable rivers—with an outlet to the Atlantic and accessible from thousands of rich, rolling farms—it had become the wealthiest metropolis in America and the most culturally advanced. It had a university, a library company and a philosophical society. It supported a hospital, fostered printing presses for books and magazines, and possessed a "celebrated Orrery," invented by Dr. David Rittenhouse. (This was an apparatus for measuring the movement of the planets and their satellites in the solar system.)

The wealth of the city, made from profits on apprentices, indentured servants, journeymen, farmers, and the "industrious poor," drew craftsmen to Philadelphia and increased their opportunities. There were skilled shipwrights in the shipyards on the Delaware, expert carpenters, joiners, master masons, watchmakers, clockmakers, weavers, ironmongers, and silversmiths.

All this must have dazzled the teenaged Robert Fulton. He would remain a city dweller for the rest of his life.[3] Fulton never could keep still for long. As a boy, he already showed a lifelong characteristic: nervous energy. Even when he talked, he often stood up and paced. Even when he had to remain seated, he moved his arms, head and entire body so vigorously that some listeners wondered if he were quite normal. Indeed, he must never have appeared commonplace. He had developed early a talent—eventually to become an art—for meeting the prominent people wherever he went. For example, he had met Ben Franklin somehow and that genial promoter of youth-with-ability had introduced him to many of the other worthies of Philadelphia.

Even as a youth, "Quicksilver Bob" must have looked striking, giving promise of his eventual height which would exceed six feet. Not yet as heavy as he would become, he had a slender, graceful figure. He wore his full, curly, dark brown hair carelessly scattered over his forehead and falling around his neck. His complexion was fair, his forehead high, his eyes dark and penetrating. His brow was thick; his mouth and lips, full but well proportioned, gave him the appearance of a talker.

He would never have acquired the underdog look so easy for apprentices to assume. Somewhere, somehow, he early became so self-confident that little could faze him. His mother doted on him. He awed his sisters.

He must even have impressed his master, Jeremiah Andrews. Instead of contenting himself with the usual menial tasks of an apprentice, Fulton soon developed a specialty in the jeweler's trade. A custom then was to commemorate the dead by wearing rings or lockets decorated with the hair of the departed woven in complex designs. Andrews advertised "hair worked in the neatest manner." Young Fulton evidently did this work well.

From this he progressed logically to a related specialty. Often the hair designs were on one side of the locket and a symbolic water color of a tomb (or something similar) on the other. He began doing such tiny paintings. From this, he advanced to execute portraits in miniature. These appeared on chips of ivory, usually worn in lockets. This proved to be a lucrative sales item for Andrews. Gradually, Fulton took over this portion of the business. White's *Directory of the City of Philadelphia* for 1785 lists "Fulton, Robert: miniature painter, Corner of 2nd and Walnut Streets." This is the first appearance of his name in any record. The address is the same as that for Andrews.

The jeweler apparently had no talent for painting. The leading miniaturists of that time in Philadelphia were James and Charles Willson Peale. They may have given Fulton lessons. Yet it's more likely that he copied their work and taught himself. (Fulton never mentions such teaching, and no contemporaries suggested it.)

The earliest serious picture that can be attributed to Fulton is a pen and ink drawing done in 1783, a reproduction of a French print. James T. Flexner, the art critic and biographer, says that it shows

> little but care in copying. Mutually contradictory miniatures have been attributed to his professional period in Philadelphia; the pair of portraits of Mr. and Mrs. John Kittera are typical of the group that is most likely to be authentic. Unsophisticated in the extreme, showing the naive freshness of self-taught artisan painters, they fall into the category commonly known as "American primitives." The

drawing, although the result of aching care, is poor. Some parts of
the faces are flat; other features, such as the ends of the noses and
the points of the chins, have been modeled too heavily and stand
out from the ivory-like little hard round balls. Depicted with a
network of clumsily crossed lines, Mrs. Kittera's hair resembles a
bird's nest perched on her head. The coloring is neither bright nor
very harmonious.[4]

Although his miniature paintings show conscientious work, they
don't show genius, just promising talent. In that era of America's
history, a portrait painter held a commercial and artistic position
similar to what a portrait photographer holds now. Only the truly
talented, like the Peales, achieved much fame or even signed their
work. Therefore, much of Fulton's work in Philadelphia is lost or
unidentified and cannot be definitely attributed to him.[5]

The ambitious youth labored hard, continuing to do jeweler's
work and anything else that turned up. He still painted an occasional sign, made drawings of buildings and machinery.

It's not surprising that many artists have also been mechanically
inclined—from Leonardo Da Vinci to Samuel F. B. Morse. Mechanics and artists both need the observant eye, the sense of form,
and the creative spark. Nor did the twentieth century's sense of
separation between mechanics and artists exist in the eighteenth
century and earlier. Two hundred years ago, no one was surprised that a craftsman could also draw. The words, artisan and
artist, were once considered synonymous. Workmen owned their
means of production and their small shops proved hot beds of invention and creativity. The jeweler first drew a picture for the
customer of the brooch or locket he proposed to make. Artistic
and mechanical abilities had to go together.

Yet few craftsmen ever made as much money as Franklin did—
enough to retire early and devote himself to learning and national
affairs. Certainly Fulton did not. In 1785, he was probably still
technically an apprentice, earning little. Late in that year or early
in 1786, he suffered a double financial blow: he fell sick; also his
mother announced that she wanted to move to the western frontier, to Washington, Pennsylvania, where she could be near her
brother, the Reverend Joseph Smith; she expected her gifted son
to help her finance the move.

Young Fulton had developed a racking cough. Soon he found
himself showing "other symptoms indicating a disposition to pul-

monary complaints." [6] Today it would be diagnosed as pneumonia or tuberculosis. His illness must have interfered with his work and earning power; in those days of poor sanitation, no antibiotics, and poor medical practices, the wonder is that Fulton even survived.

Doctors prescribed red bark, laudanum, and opium; they applied blisters and clysters, measured out vomits and cathartics, and bled their patients for fevers and almost anything else. People did not call for the doctor unless they were desperate. Home remedies included rhubarb and senna, castor oil, Daffy's Elixir, tea made of quashey root or nettles, and plasters made of honey and flour, onion, garlic, and deer fat.

Fulton probably chose the wisest course; he followed the advice of friends and decided to take the waters at a fashionable watering place in Bath, Virginia (now Berkeley Springs, West Virginia).

He needed money to do that and to set up his mother and family in western Pennsylvania. He may have saved some, but certainly not enough to meet such unprecedented needs. After he had recovered his health sufficiently to travel, he met the problem with a solution he would use many times later during his lifetime: He borrowed the necessary funds. Few details of his transactions survive. Documents indicate he borrowed part, if not all, of the £80 ($3,200 today) needed to buy a farm for his mother in Hopewell Township of Washington County. For decades a family argument persisted as to who should repay it. [7]

Joseph Smith may not have been entirely disinterested in urging his widowed sister to come live near him. He was land poor, having purchased three hundred and seventy-six acres in Hopewell Township in May, 1785. He bought it on credit for £1,625 ($65,000), hoping that his prospective income of £150 ($6,000) per year as a pastor and his earnings from the farm would help pay for it. His calculations went awry for two reasons: His parishioners were frequently in arrears on his salary because transportation difficulties made it nearly impossible for them to realize any cash income from their crops and he had the same traffic troubles with his own crops. Smith had sold eighty-four acres of his land to a Thomas Polke, [8] but the buyer had similar difficulties about cash. The land was resold to Fulton who somehow borrowed the cash, which may have helped save his uncle from foreclosure.

Fulton also borrowed enough to buy building lots in Washington for his sisters and brother, agreeing to pay $5.50 per year for them "in perpetuity," another deal that haunted him later. Polly Fulton received one of the lots. By this time, she had married David Morris, a nephew of Benjamin West, the painter. The Morrises [9] became proprietors of a locally famous hostelry, the Globe. (She was to die, even before Robert, in 1815.)

A second lot went to Isabella Fulton Cooke who was widowed early. Her grave is still marked in a Methodist Cemetery in Claysville, Pa., near Washington. She had one daughter, an artist in her own right, who married into the artistic Wythe family of Missouri.

Another lot may have gone to brother Abraham Smith Fulton, a shadowy figure. The only records concerning him that have surfaced involve some exasperated correspondence Fulton was to have with him years later concerning real estate in Louisville. Abraham Fulton wanted Robert to finance him, but he didn't give enough information, including scaled drawings of the property, to satisfy his inventor brother.

Yet another lot went to Elizabeth Fulton who had married a man named Scott, but was already a childless widow in 1786. She sold her property in Washington and stayed with her mother on the farm until the matriarch of the family died in 1799. Robert willed the farm to her upon his death.

One suspects that Fulton swashbuckled a bit for the benefit of his relatives as he signed the deed dated May 6, 1786. An indication of his self-assurance lies in the fact that he agreed to take on such commitments even before he had reached his twenty-first birthday. The only financial prospects that lay before the young man, who styled himself "yeoman and miniature painter of the city of Philadelphia," were debts.

The farm is now part of a larger one owned by a family unconnected with the Fultons. It lies in a pleasantly wooded and rolling countryside on Cross Creek. Foundations that have been uncovered by plowmen indicate that the Fulton log cabin stood at the crest of a hill, with a good view toward the northeast. The route that passes nearby has been renamed Fulton Road.

But Washington County was poor and the problems in transportation made farming unprofitable. Joseph Smith's crisis was not ended by Fulton's help. In 1787 his salary as pastor was three years past due. He again could not meet payments when due on his farm. His land, his improvements upon it, and his hopes of remaining among his people stood in danger of forfeit. The people were called together and the case laid before them. Counsel from on high was sought. Plan after plan was proposed and abandoned. The congregations could not pay him because no one had any money.

In despair, they adjourned to meet again the following week. In that period, they found a Mr. Moore who would grind their wheat, the only remotely saleable crop they had, at a reasonable cost. They took the wheat to Moore's mill. Some gave fifty bushels, some more. Soon the flour was ready to go to market. Again the people were called together. After an earnest prayer, the question came: "Who will run the flour to New Orleans?"

No one volunteered because the trip was perilous. Months must pass before the adventurer could hope to return, even though his journey should be fortunate. Wilderness lay along nearly all the way, and the Indians' treachery was a byword. More than once, a boat's crew had gone and never returned.

At length a hoary-headed man, William Smiley, an elder in the church, sixty-four years of age, arose and to the astonishment of the assembly said: "Here am I; send me."

The old man's volunteering shamed two of the younger men into agreeing to accompany him. A day of departure was set, the people of the congregations met at the riverbank at Charleston (now Wellsburg, W. Va.), a hymn was sung, a prayer was offered by Pastor Smith; then Smiley called out, "Untie the cable and let us see what the Lord will do for us!" The cable was loosed, the keelboat poled by the three boatmen began to move away downstream.

Months passed. No word. Then one Sunday William Smiley reappeared in church. Of course he could give no report of business on the Sabbath, so everyone had to wait until Monday. On that morning, he reported, after suitable prayers of thanksgiving for his safe return, that he had sold the cargo of flour at the excellent price of twenty-seven dollars a barrel. Then Smiley emptied a large purse before the astonished congregation, many of whom saw more hard cash at this one time than they had seen altogether in their lives before.

There was enough to pay Pastor Smith all his arrearages and even for another year in advance! Each young man who had accompanied Smiley was awarded one hundred dollars; Smiley himself got three hundred, and each donator of wheat won a dividend.[10]

A tale like this would have exhilarated Robert Fulton and implanted in his mind the dramatic need for changes in transportation in this new, vast country.

3 CONVALESCENCE IN BATH

From the rustic life on the western frontier in Pennsylvania, young Robert Fulton traveled east about 160 miles to a dramatically more sophisticated world at Bath then on the Virginia (now West Virginia) side of the Potomac River across from Maryland. He took the waters there for only a short time in May, 1786, but the visit did far more than improve his health. It changed his life by convincing him to make a career decision that he had been flirting with for several years.

The gentry of Virginia, Maryland, and Pennsylvania convalesced and vacationed at Bath. George Washington stopped there many times en route to or from visits to his extensive properties in the western areas of Virginia and Pennsylvania.[1] In Pennsylvania up until this time, Fulton had dubbed himself yeoman. Then he had added "and miniature painter." In Bath, he dropped the yeoman and became only "miniature painter."

"In the unrestrained intercourse of a watering place," Fulton found himself on equal footing for the first time with men in fine linen shirts who had been to Europe. With his fire of excitement, energy and eloquence, he impressed them, as he would impress almost everyone he met for the rest of his life. By this time, he had learned usually to control the flash and flare of his language and ideas. Never underestimating his previous achievements and never letting facts cloud an impression he wished to make, he bemused nearly everyone in his presence. Only the incorrigible skeptics or the unusually perceptive listeners had doubts about him—and these usually arose only later when they were away from his compelling personality.

The visiting gentlemen, of course, considered themselves connoisseurs of art. They praised him and his work lavishly, firming the ambition he already had to go to Europe to study painting.

As he moved freely and gracefully in this exalted company, their example firmed still another ambition—to become rich and a gentleman. The order in which he achieved those goals did not matter to him, but he recognized the interrelationship between the two. In the eighteenth century, America was still a class society, modeled after that in Great Britain, although in diluted form. While a Franklin could move up in society from tradesman to savant, he was the exception. He had earned enough as a printer to retire. Young Fulton may have painted wealthy people in Philadelphia, but he still entered their homes, figuratively, through the tradesman's door.

For example, he did a pastel portrait of Clementia Ross, sixteen-year-old daughter of John Ross, a Philadelphia merchant to whom Franklin had given him a letter of introduction. Years later, in 1808 or 1809, Fulton attended a Birthright Ball while revisiting the city. Clementia Ross's daughter was present and Fulton sat by her. "While I was unknown and friendless," he said, "I took a likeness in crayon of her [your mother]; a beautiful young girl." [2] The daughter did not know that the then-famous inventor had done her mother's portrait because the mother had never thought it important enough to remember and recount.

The artist must have chafed at his tradesman's position and longed to improve it. At Bath, he found evidence he could do so if he were to become a skilled artist.

At Bath, he also must have learned something about his physical and mental health. The poverty of his youth had left a tendency toward respiratory ailments. He had had pneumonia, and would eventually die prematurely from it. He also suffered from psychosomatic illnesses, not surprising in one so high strung. These took the form of periodic bouts of listlessness, low fever, and despondency. He was already learning the cure: new projects, new interests, new friends. For the rest of his life, but especially for the next two decades, he would be often moving on to new adventures.

At Bath, Fulton may have met a man who, although twenty-two years older than he, was remarkably like him. If he didn't meet him, he certainly heard about him because James Rumsey had captured the imagination of the people who lived along the Potomac River. He had been engaged by General Washington and his associates to make the river navigable from headwaters to mouth by dredging and canals. This was the most advanced engineering project yet tried in the United States and created an excitement comparable to that when the Panama Canal was dug more than a century later.

Rumsey had begun this project in mid-1785 and was still sweating and cursing over it when Fulton arrived at Bath the following summer. Rumsey kept his headquarters at the watering place in his inn, the Liberty Pole & Flag, which his long-suffering wife was wisely maintaining. (From experience, she knew what happened to most of his engineering projects—failure.) Although this project didn't completely succeed, it did have its triumphs—which seem especially astonishing when considered in the light of the fact that Rumsey had never seen a canal or locks before he proceeded to build them. His locks at the Great Falls were considered among the great engineering feats of the eighteenth century and were commented upon by every scientific publication in Europe and the Americas. The descent was 76 feet in 1,250 yards. High cliffs rose directly from the water below the falls, which made the return to the river difficult. The last two locks were cut in solid rock.

In those days, engineers had no means but hand drilling to make the powder holes in the stone and nothing but old fashioned black powder to do the blasting or "blowing" as it was called then. All earth had to be hand dug with plow, pick and shovel and hauled away in wheelbarrows or horse-drawn wagons.

Blasting methods were dangerously crude. After the powder hole had been laboriously drilled, it was poured nearly full of coarse blasting powder, and the mouth was then tamped with clay, leaving a small orifice into which some fine powder was poured for priming. The fuse was a strip of brown wrapping paper soaked in saltpeter.

In 1786, Rumsey made this report to Washington and associates, saying he had lost two blowers: "One run off, the other blown up; we therefore was Obliged to have two new Hands put to Blowing and there was much attention gave to them least Axedents should happen." [3]

Like Fulton, Rumsey was exceptionally good looking—large

mouth bent into an extreme Cupid's bow, handsome, almond-shaped eyes under a broad brow, a big but well-molded nose with fine nostrils. Although so courtly and handsome, Rumsey escaped effeminacy, as did Fulton, by a basic ruggedness.

Like Fulton, Rumsey impressed nearly everyone he met. Like Fulton, Rumsey had marked mechanical abilities. Although Fulton probably did not know it because the older man was unusually secretive about his inventions, Rumsey was experimenting with self-propelled boats. While the evidence is not water-tight that the two met at Bath, the likelihood is high. It is certain that Robert knew Rumsey well years later when they both lived in London.

Rumsey was born in Bohemia Manor, Cecil County, Maryland in March, 1743. He had little formal schooling, but was an ingenious craftsman, having become both a blacksmith and millwright. Some documentation survives to indicate that he worked with a jet-propulsion idea using steam in January, 1785. On March 10, 1785, he wrote a letter to Washington that shows that Rumsey knew the advantages steam could bring. Even earlier, in November, 1784, he had mentioned steam as a method of propulsion. However, he was thinking in theory and had made no model using steam by 1786.

When Fulton stopped in Bath, Rumsey was experimenting with a model of a pole boat or stream boat. It was two miniature boats, with poles reaching downward over the gunwales, joined together side by side with a paddle wheel between them. When the device was in a current, the water spun the paddle wheel which actuated the poles. The poles pushed on the bottom to move the boat forward. Rumsey had already demonstrated the contraption to Washington who had been amazed to see the little model walk upstream.

The General gave Rumsey a certificate:

> I have seen the model of Mr. Rumsey's boats, constructed to work against the stream; have examined the powers upon which it acts; and have been an eye-witness to an actual experiment in running waters of some rapidity, and give it as my opinion (although I had little faith before) that he has discovered the art of propelling boats by mechanism and small manual assistance against rapid currents; that the discovery is of vast importance, maybe of the greatest usefulness in our inland navigation, and if it succeeds, of which I have no doubt, that the value of it is greatly enhanced by the simplicity

of the works which, when seen and explained to, may be executed by the most common mechanic.[4]

Supported by such a statement from the esteemed Washington, Rumsey won from Virginia a ten-year monopoly. Maryland soon gave him the same. More cautious, South Carolina resolved that if the boat proved out, he should be rewarded.

With Washington's vague words, Rumsey succeeded in giving the impression that his stream boat was a steamboat. If the mistake was accidental, Rumsey made no effort to correct it. That's why many histories erroneously report that he invented a model steamboat in 1784. He built a streamboat at that time.

There's no way to determine now whether or not Fulton and Rumsey [5] discussed steamboats in 1786 when Robert visited Bath. Fulton never said that they did, but he never admitted that anyone influenced him.

Yet, even if Rumsey mentioned his experiments, it's unlikely that the secretive inventor revealed any significant details. Furthermore, the streamboat soon proved itself impractical. It could go against the current only. Rumsey would abandon it and the Potomac project and turn to steamboats. He was already catching the steam fever which would consume so many—even Fulton.

But in Bath, art and health dominated the young man's thoughts. He had to regain the latter to practice the former. He tackled his immediate two goals with customary single-mindedness. By about the first of June in 1786, he was sufficiently well to return to Philadelphia to set himself up as a freelance artist.

He moved from Andrews' shop to a place of his own, advertising on June 6, 1786: "Robert Fulton, miniature painter and hair worker, is removed from the northeast corner of Walnut and Second Streets to the west side of Front Street, one door above Pine Street, Philadelphia." [6]

The address was one short block from the Delaware, within sight of another experiment in self-propelled boats—this time a real steamboat built by a wild-eyed, scruffy genius, John Fitch. Again, Fulton never mentions the Fitch vessel, but it would be incomprehensible that he had never noticed it. All the idlers and curious of Philadelphia had watched the strange vehicle take shape.

Fitch [7] was born in January, 1743, two months before Rumsey,

in Windsor, Connecticut. He received only six years of common
schooling. Then he went to work, first on a farm and later apprenticed to a brass founder and clockmaker. Unlike Fulton's, his
apprenticeship was a disaster, his master teaching him little and
mistreating him physically. He ran away, tried to set up his own
business, but this failed. He served as a gunsmith in Washington's
army, a relatively successful and happy period in his life. After
the war, he went adventuring in the West and was captured by Indians. He escaped and settled for a while in Bucks County, Pennsylvania in 1785, to run a cider press. He may have met William
Henry at this time. Somehow, he caught the steamboat fever,
perhaps from Henry.

Henry, Rumsey and Fitch were all inlanders who nevertheless
experimented with water craft. The inland areas of American suffered from acute transportation problems at that time and inlanders were straining their creative faculties to solve the difficulties. People living along seacoasts could get around more easily by
sail. Sailboats did not perform well on many of America's inland
rivers.

Fitch first independently hit upon the idea of an atmospheric
steam engine in 1785, unaware that Newcomen had already invented it. He had then thought of a land steam vehicle, but abandoned this because the roads were so bad and no iron mill in
America was then capable of making rails for a railroad. He logically turned to the idea of a steamboat.

His first thought was for a continuing chain of paddles, but he
also thought of the paddle wheel. He made models of both versions and went to Philadelphia to find backers. Particularly, he
tried to interest the octogenarian Ben Franklin. As usual, Franklin
listened with a show of interest and support, but he was old and
considered himself an expert on steamboats, thus tending to shut
his mind to Fitch's ideas.

On December 2, 1785, Franklin's paper on steamboats had
been read for him before the American Philosophical Society.
The paper contains more than a dozen notions on boats which
had entered his mind during a boring ship voyage when returning from France earlier that year, months before he had heard of
Fitch.

Concerning paddle wheels, the sage wrote: "This method,
though frequently tried, has never been found so effectual as to

encourage a continuance of the practice." The failure, he continued, was due to a waste of energy; only the bottom paddle drove the boat forward, for the one entering the water tended to lift it up and the one emerging to pull it down.

Next, Franklin revived the suggestion made in 1753 by the Frenchman, Daniel Bernoulli, that if a stream of water were driven out of the stern of a boat below the water line, its reaction on the body of water in which the boat floated would drive the boat forward.

Bernoulli had won the Academie des Sciences competition in Paris by proposing the use of a vertical tube, with a funnel at the top, into which buckets of water (dipped or pumped from the side) could be poured. This water, Franklin theorized, could be drawn in at the bow and driven out at the stern. He concluded: "A fire engine might in some cases be applied to this operation with advantage."

Franklin's rambling description of jet propulsion appealed to eighteenth century Americans because they had not mastered the trick of changing the reciprocating motion of a piston into the rotary motion that would be needed to drive a wheel or a chain of paddles.

With the technology available in the eighteenth century, jet propulsion was impractical, but the appeal of the idea and Franklin's prestigious support of it gave the notion wide credence. Rumsey, Fitch and others wasted time on it before they finally dropped it.

Fitch and most other Americans had trouble with money. There was an acute shortage of liquid capital in colonial America and in the early United States. Britain had prohibited the formation of financial institutions in the colonies under her rule. When independence came, the problem was not magically solved because most wealth still lay in the form of land. Even men rich in land often had trouble raising cash. Furthermore, the currency depreciation that had occurred during the Revolution made the cash situation even tighter. This is fundamentally why all the early steamboat inventors—even Fulton—had so much trouble over money, although all but Fitch had connections with wealthy men, or were themselves wealthy.

Fitch failed to obtain subsidies for his steamboat from the Continental Congress, then applied to several state governments, who refused funds but granted him exclusive rights to operate steamboats on their waters. Rumsey had succeeded in doing the same thing in Maryland and Virginia. Fitch and Rumsey thus paved the way for Fulton to come along later and win state monopolies.

Finally, Fitch solved his money problems, after a fashion, by forming a joint stock company in 1786. Such companies had been set up increasingly often from about 1770 as a means to raise capital.

Tench Coxe, a Fitch acquaintance, had organized the Pennsylvania Society for the Encouragement of Manufactures and the Useful Arts, which in spite of its benevolent title was a joint-stock manufacturing concern whose object was to produce the greatest possible quantity of cotton at the lowest possible wages.

Fitch followed his example and formed a company that issued forty shares. Half of these were to belong to the inventor for his patent and for the labor he would expend to bring it to perfection. The twenty remaining shares were sold at twenty dollars each. (Although he sold them easily, he had trouble collecting and eventually had to settle for three hundred dollars, three-fourths of the amount pledged.)

Early in 1786 he put his rowing apparatus into a small skiff, using a three-inch cylinder engine. The thing failed miserably, to the amusement and jeers of the loafers watching on shore. Fitch made modifications and tried it again toward the end of July, 1786. As soon as the steam was up, the piston began to churn, the axletree to rattle and the oars on the side to cleave the water with strong, regular strokes. The skiff moved against the current of the Delaware.

In this version of the rowing apparatus, the lower arms of the paddles were attached to cranks and the upper arms ran through holes in the frame above to keep them in place. When the boat was in motion the sound of the wooden handles knocking against the holes through the frame made a frenzied rattling. It also caused much friction. Fitch remedied this with arms acting on gudgeons.

Columbian Magazine described the proposal for a full-scale boat this way (based on the experiment with the skiff and three-inch cylinder):

> The steam engine is to be similar to the late improved steam engines in Europe. The cylinder is to be horizontal and the steam to work with equal force at each end thereof. The mode of forming

a vacuum is believed to be entirely new; also of letting water into it, and of letting it off against the atmosphere without any friction. The undertakers are also of opinion that their engine will work with an equal force to those later improved engines, it being a 12-inch cylinder. They expect it will move with a clear force, after deducting friction, of between 1100 and 1200 pounds weight; which force is to be applied to the turning of the axletree on a wheel of 18 inches diameter. The piston is to move about 3 feet and each vibration of the piston turns the axletree about two-thirds round. They propose to make the piston to strike 30 strokes in a minute, which will give the axletree about 40 revolutions. Each revolution of the axletree moves 12 oars 5½ feet. As six oars come out of the water, six more enter the water; which makes a stroke of about 11 feet each revolution. The oars work perpendicularly, and make a stroke similar to the paddle of a canoe. The cranks of the axletree act upon the oar about one-third of their length from their lower end; on which part of the oar the whole force of the axletree is applied. The engine is placed in about two-thirds of the boat, and both the action and reaction of the piston operate to turn the axletree the same way.[8]

The ship as described was built by Brooke & Wilson, ironmongers and shipbuilders, but Fitch, with the aid of Henry Voight, a Dutch craftsman and watchmaker with a shop not far from Fulton's address, built the engine himself to save money. The shipbuilders had completed the boat (which looked like a small, flat-decked barge) by April, 1787, but the engine remained unfinished until mid-June.

Boat, rowing apparatus, furnace, boiler and cylinder were ready for a trial run on August 22, 1787.

We don't know if Fulton actually saw this experiment because he left for London sometime during that summer. (He would never have admitted viewing it even if he had.) However, proof that he was familiar with Fitch's experiments, at least later, lies in the fact that he copied in his own hand twelve pages from Fitch's diary for 1790 dealing with later versions of the steamboat and engine.[9]

Some Constitutional Convention delegates were cajoled aboard the vessel at the Front Street wharf as it "walked the waters like a thing of life." Although the machinery was still imperfect, it bucked the fast current of the Delaware at two and a half miles an hour.

Rembrandt Peale, the painter of many of the delegates, described the boat this way:

> On the deck was a small furnace, and machinery connected with coupling crank projecting over the stern to give motion to three or four paddles, resembling snow shovels, which hung into the water. When all was ready and the power of the steam was made to act, by means of which I was then ignorant, knowing nothing of the piston except in the common pump, the paddles began to work, pressing against the water backward as they rose, and the boat to my great delight moved against the tide, without wind or hand . . .[10]

Like Rumsey, Fitch would also play a later role in Fulton's life. It's unknown whether they ever personally met; if they had, they would not have been friends, as Rumsey and Fulton became, because of the personality differences between the two. Fitch was a cantankerous, irascible man, already well along the road to the alcoholism which would eventually kill him. Fulton tried to keep himself under tight control and usually succeeded. Like most people uncertain that they can always keep a good rein upon themselves, he distrusted—even abhorred—those such as Fitch, who made no pretence at personal restraint.

While Fitch spent the rest of the 1787 summer giving exhibition runs up and down the Delaware and into the Schuylkill, Fulton made himself ready for London. The Fitch vessel needed improvements; so did Fulton's artistic abilities.

So persuasive was Fulton that he had relatively little trouble in raising funds to go to London to study painting. If this seems surprising, remember that sending promising youngsters off to Europe for study had become an American tradition. Franklin had been so sponsored years before. Benjamin West had been sent off, too, and he was now one of the world's great painters.

Robert borrowed forty guineas (the principal patron one Samuel Scorbitt) and set sail late in the summer of 1787. Even he didn't know he would be gone nearly twenty years; he had expected to study only for a year or so. One marvels at the self-confidence of a youngster not yet twenty-two years old setting off with less than three hundred dollars and with no certainty about how he would earn his living! What he lacked in tangible assets, he made up for in the intangible, notably in a certainty in his own abilities that would never leave him.

4 ARTIST IN ENGLAND

Robert Fulton made a painful discovery soon after he had arrived in London. "Painting," he wrote his mother, "Requires more studdy than I at first imagened in Consequence of which I shall be obliged to Stay here some time longer than I expected." [1]

This admission is unique—the only instance that survives to indicate that. "Quicksilver Bob" was less than confident. One can sympathize. What had passed for artistic skill in Philadelphia showed up as provincial when contrasted to London's flourishing art, in one of its great periods of creativity.

Up until about 1700, most of the best painters working in Britain had been foreigners; and wealthy patrons continued to buy or commission foreign rather than British pictures. In 1711 Sir Godfrey Kneller, famous as a portrait painter at the end of the seventeenth century, started an academy in London for training British artists; and the St. Martin's Lane Academy was revived by Hogarth in the 1730s.

But it remained difficult for painters to find a market for their work. Hogarth started the idea of putting paintings on show. However, it was not until the Society of Artists was founded in 1760 that artists had a regular means of exhibiting their paintings. Hogarth evolved a completely new style of painting designed to appeal to middle-class people. His portraits were primarily likenesses, and he introduced the informal group portrait known as the conversation piece—the most popular of all eighteenth century genres.

Equally important as an influence on British painting was Sir Joshua Reynolds. He came back to England in 1753 after studying

for several years in Italy, and established himself as the only rival
to the Scottish artist, Allan Ramsay, then the leading portraitist in
London. Reynolds was as determined as Hogarth to raise the
status of the British painter. As first president of the Royal
Academy, founded in 1768, he delivered a series of annual dis-
courses to the students, in which he taught that history painting
was the most important genre.

A Reynolds intimate was Benjamin West who followed the advice of his friend, to become the history painter to George III
from 1772 to 1801. He was also a charter member and president
of the Royal Academy from 1792 to 1802. He had left America
for Italy in 1756, never to return, although he remained loyal to
America. In 1763, West moved from Italy to London. Besides
Reynolds, he soon came to know well other great English artists,
such as Thomas Gainsborough.

West gained widespread popularity, probably more than he de-
served as a painter. His "Agrippina With the Ashes of Ger-
manicus" was admired by the King. One of his best known works,
"The Death of Wolfe," popularized the use of modern costume in
historical paintings. In 1802, he visited Paris and exhibited his
final sketch for "Death on a Pale Horse," which anticipated development in French romantic painting.

Fulton's chief asset upon arriving in London was a letter of introduction to West. Many aspiring American artists had been so
introduced, and West treated them all with kindness.

Robert would later describe West to his brother-in-law, David
Morris:

> Your Uncle West is now at the head of his proffession and Presides
> at the Royal Academy over all the Painters in England—But he is a
> Great Genius and merits all the honour he has obtained—he has
> stedily persued his Course and Step by Step at length Reached the
> Summit where he now looks Round on the beauties of his In-
> dustry—an Ornament to Society and Stimulis to young Men.[2]

The letter gives an idea of why Fulton got along so well with the
great—and yet, he himself impressed almost everybody. An acquaintance named George Sanderson took the trouble to write to
Fulton's mother on July 24, 1788 when he returned to America.
He praised Fulton's progress "in the liberal Art of Painting" and
also mentioned the influential friends that "his personal accomplishments and prudent behaviour" had won for him.

West had put Fulton to work at learning the sophisticated modes of English painting. The young man became one of many who, in the morning, brought their canvases to the master for criticism—waiting in rows on benches supplied for that purpose.

The young American did not stand out as an artist among the neophytes. His homesicknesses and brave show of confidence come through in many passages in letters to his mother.

> All things work together for good in the end and I am Convinced my exertions will have a good tendency . . . There is nothing Interupts my happiness here but the desire of seeing my Relations but time will bring us together.[3]

At another time, he wrote, "Pleasure to hear of Abrahams attention to you tho I am sorry he has run away with the Idea of [my] Getting Rich—I only wish it was true—but I Cannot Conceive from whence the Report arose." [4]

The report arose from Fulton himself who had fallen into habits of dropping such hints and innuendoes. Perhaps this practice helped him offset the miseries of reality. Sometimes he didn't know where his next meal would come from. He wrote his mother,

> Many Many a Silant solitary hour have I spent, in the unnerved Studdy Anxiously pondering how to make funds to support me till the fruits of my labours should sifficant [?] to repay them.[5]

He was often so hard up that he could scarcely afford postage. In one letter dated July 31, 1789 he asked his mother

> . . . to write small and close that you may say a great deal in small cumpas for the ships often put the letters ashore at the first port they make. They then come post to London and I have often paid half a guinea for a small package of letters—the better to accomplish this you better buy letter paper as it is thin as we pay according to the weight and not the size so if you can send me a pound of news upon an ounce of paper I shall save almost a guinea by it.[6]

In those days the recipient, not the sender, paid the postage.

Consider also the effect of London upon a young provincial whose previous standards of sophistication and urbanity lay in

Philadelphia. Although the City of Brotherly Love boasted a population of 43,000, it was a village compared with the teeming
million of London, the biggest city in the world.

The streets were ill lit, cobbled, filthy, evil smelling and un-
drained. They had no foot paths and no protection for pedes-
trians except periodic iron posts along which the walker had to
grope his way. The prosperous traveled by river if at all possible,
or used sedan chairs if they had to go by land. Fulton seldom
could afford either.

Ideally, he would have preferred to live with West, if only to
minimize the difficulties and disagreeableness of travel in London. But West did not extend an invitation, even to a fellow
townsman from Lancaster, Pennsylvania. Perhaps the West household was already full or possibly he could see little talent in the
youngster. Another American student, Gilbert Stuart, did live
with West for a time.

Robert finessed the question of why he did not live with the
master with this explanation to his mother: "Mr. West and me are
on a very familiar footing and when he is in town pays me much
attention which is extremely agreeable as we live near each
other." [7]

Leigh Hunt, who later frequented West's studio, described the
atmosphere in the picture rooms where many of the master's
large canvases hung:

> Everybody trod about in stillness, as though it were a kind of holy
> ground . . . The talk was very quiet, the neighborhood quiet, the
> servants quiet; I thought the very squirrel in the cage would make a
> greater noise anywhere else. James, the porter, a fine tall fellow
> who figured in his master's picture as an apostle, was as quiet as he
> was strong; standing for his picture had become a sort of religion
> with him. Even the butler, with his little twinkling eyes full of pleas-
> ant conceit, vented his notions of himself in half-tones and whispers
> . . . My mother and I used to go down the gallery as if we were
> treading on wool. [8]

In West's studio, they discovered "the mild and quiet artist at
work, happy for he thought himself immortal."

Royal Academy records show that Fulton changed lodgings
often, probably because of problems paying the rent. This rootlessness would have a pronounced effect upon him. It perma-

nently confirmed him in his passionate longing to be somebody—preferably a rich and cultured gentleman whose accomplishments would assure his place forever in history. It may even have determined his eventual choice of the steamboat as the object which would help him achieve his goals. The boat itself is an embodiment of a kind of practical rootlessness.

West taught Fulton a lesson that had nothing to do with painting. The older man had achieved more reverence and influence in London than his artistic skills merited because he impressed his fellow men by a conspicuous saintliness. He dressed soberly, he helped everyone, he never displayed fits of artistic temperament. And West had become wealthy.

Yet Fulton must have fretted and stewed. He won almost no commissions. He lied to his mother in 1789, claiming that "my pictures have been admitted this year into the Royal Academy."

In 1790, Fulton visited France briefly to improve "my taste and eye" [9] by studying the pictures there. He described what he saw like an engineer explaining the mechanism of some new device. Absent is the tumult of enthusiasm you would expect from a dedicated artist. He might just as well have been a mechanic discussing wage rates in a strange land, but perhaps the trip did help him.

He won a distinction in the following year which he had prematurely claimed earlier. In 1791, he showed two portraits and two subject pictures at the Society of Artists and had another pair of portraits accepted by the more prestigious Royal Academy. Yet he exaggerated even this achievement to his mother. The two pictures at the Royal Academy became eight. Fulton also said they "Recd every posable mark of Approbation that the Society could give but these exertions are all for honor—there is no prophet arising from it. It only tends to create a name that may hereafter produce business." [10]

Fulton's claims of "approbation" don't stand up. The newspapers of that period scarcely mention his paintings.

None of the original portraits of that time that can be definitely attributed to him are known today. We have only three engravings made in 1793 from Fulton's historical paintings. Two show Mary Queen of Scots under confinement and Lady Jane Grey the night before her execution. The prints are without distinction. The drawing of the hands is poor. The result is neat but not impressive. Despite their heartrending subjects, the pictures reveal little melodrama. They show a melancholy restraint—pretty but not powerful. One can only hope that the engraver did not do Fulton full justice. Flexner says:

> Gone is the sincerity, the aching search for truth which had charac-
> terized the crude miniatures we have attributed to Fulton's Phila-
> delphia period. In its place we find a superficial sophistication, a
> somewhat hesitant use of technical expedients which the artist has
> learned by rote, without feeling them at all. Fulton's style could
> now express no emotion deeper than sentimentality.[11]

Perhaps Fulton's success at being "hung" at the Royal Academy did "create a name" that would "produce business." An aristocrat, from outside London where the competition was less, commissioned a portrait of himself and introduced the painter "to all his friends." Lord William Courtenay in his country seat in Devonshire gave Fulton his chance at last. Yet Fulton's letter home sounds singularly unexhilarated:

> I was invited by Lord Courtney down to his Country seat to paint a
> picture of him which gave his Lordship so much pleasure that he
> has introduced me to all his Friends—And it is but just now that I
> am beginning to get a little money and pay some debtt which I was
> obliged to contract so I hope in about 6 months to be clear with the
> world or in other words out of debt and then start fair to Make all I
> can.[12]

For much of his life, Fulton could almost claim to be a professional guest. For long periods in America, Britain and France, he lived with, or for extended periods visited, various friends. Robert Fulton was a charming guest—amusing, vivacious and entertaining. Visiting was more the custom in those days when inns were few and of poor quality and social intercourse was the entertainment. He also cultivated the art to save money. It's probable that he even gave some of his paintings to his hosts, in lieu of payment for board and room. This may account for the fact that so few of his original paintings survive. Perhaps the recipients put them in obscure corners of their homes or even in their attics and they mouldered away or were lost.

Fulton was serious about the rules and ethics of hospitality. Although he stayed with Courtenay only a short time and actually saw more of the steward than of the baron—who lived in quasi-royal seclusion—Fulton was one of the few in America who offered Courtenay hospitality when he had to flee there from some disgrace at home.

Fulton had also raised the art of borrowing to a science. Basically, he relied on his charm to extract money and he judiciously selected the right sum. He never asked for more than he thought the lendor could lend without serious consequences if it were not repaid quickly. This accounts for the fact that he sometimes did not have to repay debts for years.

When pressed, however, Fulton did repay them—or try to get someone else to do so. For example, in 1792 he had still not paid for the building lots purchased in 1786 in Washington, Pennsylvania. He wrote to Mrs. Fulton about the troublesome problem on January 20:

> When I wrote you last I beged you would Settel everything to You[r] mind relative to the Lotts and after Regulating everything with Mr. Hoge and putting me on the way how to act I would transfer my Right in the manner you Can best Settel among yourselves—tho I could wish one of them were sold to pay Pollock—For I realy feel my honor Concerned in keeping the poor man so long out of his money nor had I the least idea of its remaining so long unpaid or I should have endeavored by some means to have it done—but I hope when I hear from you next in Answer to these letters you will have everything so Situated so I may transfer them to your wish And if no other method can be found one lott ought to be sold to pay Pol'k.[13]

In other words, settle the matter, but in such a way that it doesn't bother or cost Robert Fulton any money.

At the other end of the borrowing scale, Robert took care not to ask for trifling amounts. Increasingly now, he styled himself "gentleman"— and a gentleman did not beg a pittance. He borrowed a gentlemanly sum. By the end of his life, Fulton had grown so consummate in such practices that he could fool even himself. He could live like a lord and not realize that he was virtually bankrupt.

The visit at Powderham holds unusual significance because there he met two aristocratic friends of the Courtenays—the Duke of Bridgewater and Lord Stanhope—who led him toward a new path of thought and endeavor, invention.

When Robert Fulton met him, Francis Egerton, Duke of

Bridgewater, was already in his late fifties, a dry wisp of a man and a lifelong bachelor. He had been disappointed in love. As a young man he had been engaged to the beautiful Duchess of Hamilton. He had left London after the engagement had been broken off to devote himself for the rest of his life to the development of his estates at Worsely, near Manchester.

His property contained coal, but he couldn't transport it cheaply to Manchester and other markets. This led to his construction in 1767 of a canal to Manchester that ushered in the canal craze in England. He was at first thought a lunatic, but madmen could do almost as they pleased in England at this time—if they were also peers of the realm. The venture, reportedly earning as much as one thousand percent on the Duke's original investment, proved highly successful financially. It was built by the celebrated engineer, James Brindly, and its construction included a famous aqueduct across the Irwell River. The success led to wild speculation in canals that were projected all over England, beginning in the 1760's and 1770's.

Devonshire proved ideal for the artificial waterways, and the Duke had interested himself in projects in that county. That accounted for his visit to Powderham.

One Devonshire canal project involved still another aristocrat, the Earl of Stanhope. This scheme, projected in 1793, was intended to join the Bristol and English Channels, touching Holsworthy, Stanhope's manor.

About forty years of age when Fulton met him, Stanhope doubly impressed the younger man. First, he was a peer and one of the well-known men of his day. Fulton always gravitated to such people, like iron to a magnet. Second, Stanhope and he had common interests in mechanics.

If not a true genius, Charles, Earl of Stanhope, looked and acted like one. He was a tall, ungainly man whose long narrow face was topped by a bald head. His lean, graceless figure was always clad in simple clothes. He was married to William Pitt's sister, Hester. He advocated democracy, finding nothing inconsistent in adopting that posture while clinging tenaciously to his title and enormous, inherited landed estates. So powerful was he that his public praise for the French Revolution and later for Napoleon gave him no difficulties, despite the fact that his secretary was jailed for treason.

Stanhope was born in London in 1753 and educated at Eton and in Geneva, where a leading scientist tutored him. He painted pictures and invented a mathematical instrument. At eighteen he won a prize from the Academy of Stockholm for a paper on the pendulum. As Lord Mahon, he was elected to the House of Commons from 1780 until his succession to the peerage and the House of Lords in 1786. He opposed the war with America and later with France. Being "in a minority of one" in the House of Lords, he withdrew from parliament from 1795 to 1800.

Stanhope's inventions included a printing press and the microscopic lens which bear his name, two calculating machines, a steam carriage and a steamboat to carry coal from Newcastle to London. Critics pointed out that the latter "would have consumed its cargo before it could have reached its destination."

Much of the Earl's experimentation had the flavor of dilettantism. He flitted from interest to interest, never carrying anything to much commercial or practical success because his wealth made it unnecessary for him to do so.

He had an arrogance, too, that must have been tolerated only because it was so unconscious. For example, a letter of October 23, 1789 to Boulton and Watt, the acknowledged world leaders in manufacturing steam engines, directed the firm to send him the plans for one of Watt's best "fire engines," so that he might modify and improve them. Further, he ordered the firm to send him one of their most skilled workmen to put his plans into effect. He wrote in the third person:

> Lord Stanhope would be glad to be informed whether Mr. Boulton had made any new discovery in respect to his fire engine since his patent, and what is the nature of such discoveries.[14]

He would include them in his own contemplated improvements.

Boulton and Watt politely declined the invitation, but the fact that they even answered it showed Stanhope's exalted status.

Fulton didn't abandon art immediately. From about 1792 to 1794, he kept one foot in art and one in mechanics. But by the latter date, when he was twenty-nine years of age, he decided to change his career. Even so, it took him another two years to announce the fact to his mother. In September, 1796, Robert wrote home, "I have laid aside my panels, and have not painted a picture for more than two years, as I have little doubt but canals will answer my purpose much better."

That purpose was to win fame and fortune. Benjamin West states the reasons for the change most clearly, saying that his pupil "came to England with an intention to study painting, but doubting his success turned his attention to mechanics."

Yet it would be unfair to consider Fulton a superficial cynic who abandoned art only because he could not make enough money from it. For seven years in England, he had tried hard to become an artist. He eventually mustered the courage and objectivity to recognize that he would never be great or even outstanding.

Although he abandoned art as a career, his drafting skills would serve him well in his new engineering field. He would never paint again as a professional. He did produce some creditable portraits as an amateur when he painted only for his own and his friends' amusement.

And his long study and practice of art did something more for him. It helped educate him. It developed his sense of history, his eye for detail, his appreciation for form, and his recognition of how past and present concepts can be transmuted into something new. Art helped make him a gentleman, and it helped make him into something new—a technologist.

CANAL FEVER

One would have thought Fulton would have embarked on his new career in invention by finding a job in some mechanical capacity on the canals that had captured his imagination.

As he painted less and less between 1792 and 1794 and thus earned less and less from that source, the need for funds must have grown steadily more acute. Yet no evidence exists that he ever worked in any regular paid position. Usually, he supported himself by visiting or borrowing. Often, he borrowed from his host.

Two reasons largely account for his learning engineering by observing rather than working in it. First, he was now a "gentleman," and gentlemen did not work for others. Second, he was temperamentally unsuited to take orders from anybody. He had to run an enterprise, and as a neophyte, he was not at first technically qualified to manage any project.

One can picture Fulton forever on the move, with his luggage full of drawings, his mind full of ideas, and his conversation full of engaging notions. Depending on his audience, he could talk of mechanics, art, politics, and even philosophy. When he arrived at a destination—often the home of some cultivated man to whom he had been passed along by his previous host with a letter of introduction—he would settle in by writing letters to various personages.

While staying at Torquay, Fulton wrote to Lord Stanhope on September 30, 1793, mentioning a project for moving ships by steam and enclosing a scheme for doing away with locks in the Earl's canal. His idea was for a preponderating cistern of water to

draw the canal boat up an inclined plane from one level to another. Stanhope replied courteously enough, but informed him
that his proposal for an inclined plane had been advanced sixteen
years earlier by one Edmund Leech. However, the Earl did show
interest in the steamboat ideas because "it is a subject on which I
have made important discoveries." [1]

The "discoveries" had been borrowed from Genevois of Berne
who had proposed in 1759 that a double-ended vessel be driven
by a propeller operating like the foot of an aquatic bird. This
would work through a trunk in line with the keel and be operated
by a steam engine. Stanhope had persuaded the Navy Board in
1792 to try out his schemes, with the Earl to bear the expenses if
they proved unsuccessful. (They later did fail.)

Fulton rose quickly and lengthily to the invitation with this letter of November 4, 1793:

> . . . [to] conform with your lordship's wish I have made some
> slight drawings descriptive of my Ideas on the Subject of the steam-
> ship which I submit with diffidence to your Lordship. In June '93 I
> began the experiments on the steamship; my first design was to imi-
> tate the spring in the tail of a Salmon: for this purpose I supposed
> a large bow to be wound up by the steam engine and the collected
> force attached to the end of a paddle as in No 1 to be let off which
> would urge the vessel forward. This model I have had made of
> which No. 1 is the exact representation and I found it to spring
> forward in proportion to the strength of the bow, about 20 yards,
> but by the return of the paddle the continuity of the motion would
> be stoped. I then endeavoured to give it a circular motion which I
> effected by applying two paddles on an axis, then the boat moved
> by jerks. There was too great a space between the strokes; I then
> applied three paddles forming an equilateral triangle to which I
> gave a circular motion by winding up the bow. I then found it to
> move in a gradual and even motion 100 yards with the same bow
> which before drove it but 20 yards.
>
> No. 2 is the figure of my present model in which there are two
> equilateral triangles, one on each side of the boat acting on the
> same shaft which crosses the Boat or Ship and turns with the trian-
> gles; this, my Lord, is the line of experiment which led me to the
> triangular paddles which at first sight will convey the Idea of a
> wheel of perpendicular oars which are no longer in the water than
> they are doing execution. I have found by repeated experiment
> that three or six answer better than any other number as they do
> not counteract each other. By being hung a little above the water it

allows a short space from the delivery of one to the entrance of the other; it likewise enters the water more on a perpendicular as the dotted lines will show its situation when it enters and when it is covered the circular dots exhibit its passage through the water. Your Lordship will please to observe in the small wheel with a number of paddles A. B. C. and D. strike almost flat in the water and rise in the same situation whilst E. is the one that pulls, the others act against it which renders the purchase fruitless; while E. is urging the ship forward B.A. is pressing her into the water, and C.D. is pulling her out: but remove all the paddles except E. and she moves on in a direct line. The perpendicular triangular Paddles are supposed to be placed in a cast Iron wheel which should overhang above the water—it will answer as a fly and brace to the perpendicular oars. This Boat I have repeatedly let go and ever found her to move in a steady direction in proportion to the original purchase. With regard to the formation of ships moved by steam I have been of the opinion that they should be long, narrow and flat at bottom, with a broad keel, as a flat Vessel will not occupy so much space in the water; it consequently has not so much resistance. A letter containing your Lordship's opinion of this mode of gaining a purchase on the water and directed for me at the post office, Exeter, will much oblige your Lordship's most obedient and very humble servant.[2]

The letter mentions that June, 1793 was when Fulton first started thinking about steamboats. It is naive mechanically. The state of propulsion technology was already far ahead of this as Robert himself would discover later when he returned seriously to steamboats. The letter also shows Fulton's methodical approach to any problem: first, theorize; then put your theories on paper; finally, test your theories with models. Fulton would later add another step to the approach—change the theory when actual, full-scale vessels show flaws. For example, in his first few boats, the inventor held to the configuration of the ship's design which he here described—long, narrow and flat at the bottom, with a broad keel. But usage showed this might not be best. Therefore, he modified the design.

Stanhope didn't think much of Fulton's ideas, of course, and continued with his duck-foot propellor. He actually built a boat, which he called *Ambi-Navigator Kent,* and tried it out with a twelve hp engine built by Boulton & Watt. But it moved at only three miles per hour, and Stanhope abandoned the idea.

Canals had gripped him, as they did many in England. In the

four years ending in 1794, eighty-one Canal and Navigation Acts were obtained allowing some individual or company to build a canal. No fewer than forty-five of these were passed in 1793–1794, authorizing the expenditure of £5,300,000 (more than $200 million in today's values).

The world was crying for better transportation in the eighteenth century. Water was the only really reliable way to get from one point to another anywhere in the world. When there was no water, it was natural for people of that era to turn to some means to create a waterway.

Roads were incredibly bad, even in England which probably had among the best anywhere. Stages carrying both passengers and freight bumped along unsprung [3]; the ride was agonizing and dangerous. These were the days when a driver had to prevent his stage from capsizing on the curves by shifting his human ballast: "All lean to the right now, gentlemen, if you please!" Or, "All to the left here, thank you!" Those were the common cries of the road, while a hair-trunk or two was apt to roll off in the mire at any moment.

Fulton went with the temper of the times and turned to canals, inspecting many already operating in England, especially Bridgewater's Canal (opened in 1767) and the Grand Trunk Canal (completed in 1772).

Yet strangely, no word of this new interest surfaces in his letters home. In a 1794 letter to his brother-in-law, David Morris, he asks about London reports concerning inroads of the Indians on the frontiers, discusses European politics and solemnly predicts that "Monarchial Governments are going out of fashion." [4]

His brain teemed with ideas about waterways. He advocated small canals both because of lower costs in construction and to speed up transport of light loads. To overcome differences of levels, especially when water was scarce, he proposed to use inclined planes or vertical hoists instead of locks. Inclined planes were used a thousand years ago in China. Several had even been constructed in 1792 on the Shropshire Canal, which Fulton probably had inspected.

Fulton made improvements on the plane and took out a patent dated May 8, 1794.[5] He describes himself as "late of the City of Exeter, but now of the City of London, Gentleman." The invention's title is in the longwinded style so dear to the eighteenth century: "A machine or Engine for conveying Boats and Vessels and their Cargoes to and from the different levels in and upon Canals, without the Assistance of Locks or the other Means now known and used for that purpose." [6]

The specification runs to six printed pages, accompanied by a colored drawing. Henry Dickinson, an engineer himself from the British Science Museum and author of a biography of Fulton in 1913, commented: "It is difficult even for a trained mind to see in the specification anything more than a crude idea, ill digested; better methods, worked out in a more practical manner, were already in use." [7]

At about this time, Fulton also came up with an idea for mechanical excavation. There's no need to describe it in detail because it failed. Dickinson terms it "a crude and impracticable apparatus. The power for cutting the earth was to be obtained through the axle of the machine from the four horses employed in dragging it . . . The velocity of the flies would be so great that they would deal a shattering blow instead of the slow motion necessary with a shovel." [8]

However impractical, this is the first attempt on record anywhere to solve the problem of mechanical excavation. Fulton showed here that he possessed an essential characteristic of the successful entrepreneur—a lively sense of what the world needed. He must have soon realized its shortcomings because he never attempted to patent it, nor did he mention it in his first appearance as an author with *Treatise on the Improvement of Canal Navigation.* [9]

The *Treatise* came out in 1796. It was translated into French and Portuguese and attracted much comment. It proposed a system of small canals, wheeled boats, and inclined planes to connect major cities, even if they had to climb hills or be carried by bridges across deep valleys. The canals would be uniform and the boats of the same size so that vessels could go wherever the system went. Fulton designed various kinds of boats for different purposes— cutters for passengers, slower market boats for perishable goods, and big freighters for heavy items. He proposed operating the system under uniform rates and even a single managership.

The book contains many descriptions and seventeen drawings of mechanical devices—inclined planes, boats, aqueducts. Fulton also included calculations about speed and water resistance. This last was a genuine innovation—the first example of an inventor of marine-related devices coupling the two concepts and providing statistics to prove the connection.

Fulton's book also mixed political and idealistic theory with his technology and statistics. In a letter touting the *Treatise* to President George Washington he claimed his system of canals would "give an Agricultural Polish to every Acre of America." [10] Washington gave him a polite brush-off. He wanted nothing more to do with canals after the disastrous experience with the Potomac project and James Rumsey. The President probably didn't even read Fulton's tract which grandly claimed that canal navigation would improve internal transportation, thus making society richer and happier, spreading knowledge, removing prejudice, raising living standards and abolishing war.

Drawing a blank from Washington, Fulton sent a copy to the governor of Pennsylvania, General Mifflin, advocating that his canal ideas be put into effect in his native commonwealth. [11] Interestingly, he urged that the state undertake the project to avoid permitting any individual to profit unduly from the benefits. Mifflin showed no interest either.

The *Treatise* suffered the fault of many first works: it was long on the great results to come from the proposals, but short on the means of achieving them. William Chapman, who had helped build the London docks, rebutted the *Treatise* in 1797 with his *Observations on the Various Systems of Canal Navigation*. He emphasized how old a device the inclined plane was and how unsuitable the small canal was to a country as heavily populated as England. Chapman didn't recognize, however, the element of prophecy in the *Treatise*. Fulton described with remarkable accuracy many of the benefits that would come early in the next century with the advent of improved internal transportation—but as the result of the railroad rather than the canal.

Chapman also twitted Fulton for styling himself as "civil engineer" on the title page of the *Treatise*. True, the new author had no right to call himself that because custom had decreed that the title should be reserved for engineers who had built civil projects (in contrast to military). Fulton had nothing of this sort to his credit, yet the title was loosely used in those times and did not imply formal education in the profession.

When examining Fulton's progress between about 1792 and 1797, one could make a case that his main accomplishment had been to learn the art of existing without visible means of support. A cynic might even charge that the man was a charlatan. When it suited him, he called himself an artist, although he had painted little since 1792 and almost nothing since 1794. His canal-related projects had proven impractical. So did other devices, such as a machine for cutting marble, the excavator, a flax-spinning unit, and a new method for tanning.

He won a gold medal for the marble-cutting machine, but little else. He picked up an idea for a rope-making device suspiciously like one developed by Edmund Cartwright, the English inventor of the power loom.

Significantly, Cartwright never showed any resentment over this. He and Fulton had met sometime in the 1790's and had become fast friends. Cartwright's daughter testified that Robert's "vivacity of character and original way of thinking" made him welcome in the English inventor's home. According to her, they discussed navigating by steam interminably, because Cartwright thought that another of his inventions, the alcohol engine, could power a boat.

Undoubtedly, Fulton's chronic need for money forced him to cut corners from time to time. He could also extract funds from unlikely sources, such as the formidable Earl of Stanhope. After a grim Christmas in London in 1796, he wrote the Earl a long letter about the difficulties of introducing a canal system:

> . . . And till they are Brought About, Penury frequently Presses hard on the Projector; and this My Lord is so much my Case at this Moment, That I am now Sitting Reduced to half a Crown, Without knowing Where to obtain a shilling for some months. This my Lord is an awkward sensation to a feeling Mind, which would devote every minute to Increase the Comforts of Mankind, And Who on Looking Round sees thousands nursed in the Lap of fortune, grown to maturity And now Spending their time in the endless Maze of Idle dissipation. Thus Circumstances My Lord, would it be an Intrusion on your goodness and Philanthropy to Request the Loan of 20 guineas Which I will Return as Soon as possible, And the favour shall ever be greetfully Acknowledged By your lordship's Most obliged Robert Fulton.

In a postscript, he added, "I have also pondered much on the Liberty of Requesting a favour of your Lordship Which Realy gives me pain but My Lord Men of fortune Can have no Idea of the Cries of necessity—And I must Rely on your Lordship's Goodness." [12]

Perhaps the addendum turned the trick. Fulton got the loan, as he usually did. The letter carries a querulous tone, however, unusual for the inventor who normally presented a cheerful face to the world.

Yet sometimes Fulton needed far more than 20 guineas (worth about $800 in today's currency). For example, it cost at least £100 to obtain an English patent (about $3,600 in modern money). That's why he did not apply for many.

When he visited Manchester to study the Bridgewater Canal, he met Robert Owen there. Owen was then managing cotton mills in Manchester. (Later he became owner of the Scotch mills where he carried out his celebrated schemes in social reform.) As usual, Fulton needed money and confided in Owen who describes the request in his autobiography:

> In prosecuting an invention which had occurred to him [Fulton] for more expeditiously and cheaply digging or raising earth in forming canals . . . he had expended all his funds, and he knew not, except by disposing of part of the interest in his patent (for the inclined plane), how to obtain more, for all his means and credit were exhausted. He said there was a canal to be constructed near Gloucester, and if I [Owen] could supply him with funds to go there and see the Commissioners appointed to carry it into execution, he might perhaps succeed in obtaining a contract for digging a portion of it, and might then bring his new patent into notice and profitable action, and he would give me half the interest in the invention.[13]

Although Owen was dubious, such was Fulton's charm and persuasiveness that the social reformer eventually put up the money by buying a sixteenth share in the patent on the inclined plane. What's more, Fulton persuaded the reluctant Owen to put up money for the canal digging machine. In fact, Owen paid up when the idea for the excavator had barely taken shape in the inventor's fertile brain. The two formed a partnership in December, 1794, the main feature of which was that Owen advanced Fulton £60 (about $2,160 in today's dollars). In March, 1795, Fulton drafted a new agreement, apparently because he needed still more money:

Mr. Robert Owen having advanced the sum of £93, 8s in part towards promoting the two projects of running boats independent of locks and removing earth out of canals—it is hereby agreed that the said Robert Owen shall advance to the said Robert Fulton a further sum not exceeding £80 to enable him, the said Robert Fulton, to make a fair experiment on the earth removing apparatus; that on finishing such machines, should the said Robert Owen think proper to proceed in the partnership as per contract, he shall be at full liberty so to do. But should a partnership be presented to the said Robert Fulton previous to finishing the said machine, he shall be at liberty to accept of the same on the proposal of the said Robert Owen. And in such case, the said Robert Fulton to pay to the said Robert Owen five per cent per annum, for the monies advanced until the said Robert Fulton shall be enabled to refund the principal.[14]

Owen loaned Fulton a total of about £170 (some $6,120 in current dollars). The social reformer says that the inventor never repaid more than £60 ($2,160). We know why: neither the inclined plane nor the excavator amounted to anything. Until about 1797, Owen and Fulton had a lively correspondence, but the Englishman never heard from the other again after that time.

Was Owen bitter about Fulton? Not at all. Late in life he said:

I consider the little aid and assistance which I gave to enable him to bestow so great advantage on his country and the world as money most fortunately expended.[15]

A footnote about the excavator: In January, 1796, we find Fulton the promoter describing the digger to Stanhope as though it were a new idea in which he hoped to interest the Earl. He attempted this despite the facts that the invention's prospects were dubious and that Owen had first rights to any profits. Fortunately for all concerned, Stanhope showed no inclination to invest.

The inventor's correspondence during the 1790's often carried some prestigious return address, but with instructions to reply elsewhere, usually to what would now be called general delivery. For example, during his Manchester period when he met Owen, he dated his letters from the Bridgewater Arms in that city, but he actually lived in a boarding house at 8 Brazenose Street.

Number 8 housed some remarkable roomers, in addition to Fulton. One was Owen. Another was John Dalton who would develop the Atomic Theory; another, Samuel Taylor Coleridge, the poet. Dalton taught for a time at the Unitarian College in Manchester, and Coleridge at this period was considering joining the Unitarian Church. Fulton furthered his education in the company of these learned young men. All of about the same generation, the four proved a congenial group, talking about all manner of subjects, including canals and steamboats. It's no accident that Coleridge penned these lines at about the time he met the inventor:

> But why drives on that ship so fast
> Without a wave or wind? [16]

For the first time in his life, Fulton joined in the company of men interested in matters other than learning a profession, getting ahead, and raising money. They fired him with an enthusiasm for the French Revolution; they talked of democracy and social reform. They discussed mathematical theory, physics, and other abstract matters. The wonder is that Fulton with his meager formal education could hold his own in such company, but he did. They included him because of the fervor of his talk,[17] the tumultuousness of his ideas, and the effervescence of his moods.

To judge by his early letters, he spoke ungrammatically until he was about 30. Then, that would not have been the social handicap it is now. The eighteenth century was more casual about grammar and spelling than we are today. At about this time his grammar and spelling show improvement—again on the evidence of his letters. His manners, described as "courtly" by contemporaries, could have acquired such a veneer by the time of his Manchester days because by then he had associated with all manner of people in society, including dukes and earls.

So, picture the thirty-year-old Fulton, perhaps in Owen's rooms, drinking in new social ideas and contributing, too, with his notions on canals and engineering. A few of Owen's proposals for social reform found their way into his *Treatise*, but never into anything else in his life. How did Dalton's abstract theories strike him? Although they probably interested him, as anything new to him did, he would search for pragmatic uses because he was an eminently practical man. Dalton could have introduced Fulton to the wonders of statistics and their service. And Coleridge? While his vivacity and wide-ranging interests would have appealed to Fulton, they had little else in common.

On April 28, 1797, Fulton wrote Owen a mystifying letter from London, alluding to prospects that would end his poverty:

> . . . The agreement I have now made, I hope will crown my wishes; having sold one-fourth of my canal prospects for £1500 to a gentleman of large fortune, who is going to reside at New York. Of this £1500, I shall receive £500 [$18,000 in dollars today] on the 17th of next month, £500 in six months and £500 on my arrival in America which I hope will be about June '98 . . .
>
> In the appropriation of the first £500 it is stipulated between my partner and me, that I should go to Paris and obtain patents for the small canal system—this I calculate will cost me about £200. Of the remaining £300, I will send £60 as your portion and pay you the remainder in six months . . .[18]

(Fulton soon paid the £60 as already noted, but never anything else, although he presumably could have afforded to.)

This sale holds several mysteries. The only "canal prospects" that he could possibly sell were shares in his inclined plane. Dickinson and every other engineer who has studied the patent expresses "incomprehension" that anyone could have considered a one-fourth interest worth $54,000 in today's money.

The second mystery is the identity of the buyer—and why would the buyer want to remain anonymous, unless he had early doubts about the wisdom of his purchase? One guess is that Joshua Gilpin [19] put up the money. A wealthy Quaker from Philadelphia, he did head a group that eventually completed the Chesapeake and Delaware Canal in 1820. However, his associates did not even receive enabling legislation for the canal until 1801. The circumstantial evidence lies in the facts that Gilpin had been in England in 1796–1797, that he was interested in canals, and that Fulton knew him.

Still another mystery is what happened to the deal. Fulton never mentions it again and evidently never received the second and third installments. If Gilpin was the buyer, his memoir published in 1821 may explain what happened—his canal directors refused to adopt some unidentified proposals for canal construction because they were "untried." Some biographers have speculated that this could refer to the inventor's "canal prospects."

On the other hand, the correspondence that survives between Gilpin and Fulton makes no reference, however oblique, to a partnership on the canal system. For example, Gilpin had sent the inventor a copy of Chapman's criticism of his ideas. With his usual faculty of seeing the positive side of everything, Fulton replied to Gilpin on November 20, 1798, arguing that the critique "will tend to bring the subject to discussion and render its importance understood . . . But for the pleasure of Seeing my Canal system stand in its true Light I look to America, and to America I look for the perfecting of all my plans."

If a partnership had existed between the two, surely Fulton would have rebutted Chapman's criticism more vigorously to a presumably worried partner. If the partnership had already lapsed by this time, what harm would Fulton have seen in referring to it specifically in a private letter?

Another explanation of the partnership is possible: that it never existed, that somehow, Fulton borrowed or received £500. To explain how he got such a sum, he may have invented a reason more colorful than a loan or less disreputable than some dubious deal. Years later, after Fulton had grown famous in America, several instances occurred when he instructed his English agent to repay relatively large debts which he had "forgotten."

Perhaps it is uncharitable to view the Robert Fulton of 1792–97 as a charming confidence man. Should we rather see him during this period as struggling to find himself and his life work? In this struggle, he used the handiest tools available—his imagination, charm, and persuasiveness—to keep himself afloat. Possibly Robert Owen, Edumund Cartwright and others, whom we might think he had victimized, actually possessed the perception to recognize his dilemma and talent and the generosity to lend a discreet helping hand. Perhaps the only return on their investments they expected was to see an inventive friend develop his talents. Certainly Owen and Cartwright denied resentment.

In the spring of 1797 Fulton had enough money, from whatever source, to try his fortunes in France. Taking advantage of a temporary lull in the Napoleonic wars, he set off for Paris. He thought he would be gone long enough only to secure patents on his canal ideas. Actually, he stayed for years.

6 ON TO FRANCE

As Fulton was crossing the English Channel, a man on the ship began shouting at him in French. Fulton couldn't understand him. An attractive young Frenchwoman came to the aid of the bewildered inventor. She offered to help him with the language.

When he landed in Calais, there was trouble with his papers. He was delayed there for three weeks. While twiddling his thumbs, he learned that the Frenchwoman was delayed there as well, and was in far more serious trouble than he. Although she claimed to be Madame François, a shopkeeper's wife on her way home to her humble husband, the authorities suspected (with reason, as it turned out) that she was an aristocratic emigrée trying to slip into the country to take care of family business.

In a rare case where his heart ruled his head, Fulton rushed to the hotel room where she was under guard. The lady, who revealed to him only years later her true identity as Marie Josephine Louise de Montaut de Navailles, Duchesse de Gontaut-Biron, reported the interview in her memoirs: [1]

> "Madame François," he said, "listen to me. You are in a very bad situation and I have come to rescue you."
>
> "A thousand thanks; but be so kind as to explain."
>
> "They are going to take you to Paris and put you in prison there, and once there, you are lost. Now listen to what I have to say. Nothing could be easier than to save you from danger; nothing could be more simple; marry me; do marry me!"
>
> "Oh, thank you; but I am married already."

49

"Oh, what a pity, what a pity! I would make you rich. I am going to make my fortune in Paris."

No wonder that Robert was attracted. She had a face and figure considered classically beautiful in those days. (Nor would she be thought ugly today!) Her eyes were big and tipped in slightly at the corners to give her a gamin look. Her well-shaped mouth was generous. Her neck was long and her shoulders slanted elegantly. What's more, she was young.

She was born in 1773 and in 1793 had married Vicomte Gontaut-Biron in London where they had fled to escape the horrors of the French Revolution. Her father, Count Montault Navailles, had superintended the education of the children of Louis XVI. As a result, the entire family was on the list of proscribed emigres.

The consul from Hamburg, a Herr Schemelpeninck (who also shortly played a role in Fulton's career), eventually secured her release. The inventor's attentions at least had distracted her from the boredom and mild rigors of her imprisonment. She implied in her later memoirs that he had fallen wildly in love with her.

It's possible. In his letters, he had complained increasingly that his rootless, single state distressed him. "My sisters," he had written home, "have been extremely active in making me uncle to many, as they find my bachelor ideas still possess me. However, I am not old enough to grow musty, and possibly I may one day try how I may like it. But at present there is not the most distant prospect." [2]

With a patronizing tone, the noble lady decades afterward recalled:

> His little plan seemed to him *so simple,* and he proposed it so kindly and heartily, that while I laughed I could not help feeling grateful to him. I begged him not to trouble himself any more about me, assuring him that providence and my own good cause would be the means of saving me. He sighed and departed. [3]

A little time later, he happened to see her walking in Paris with her brother-in-law, the Marquis de Gontaut. Fulton rushed up, seized both of her hands, and, calling her Madame François, expressed delight at seeing her. Her escort informed him her name was Mademoiselle de Montault. (She was prudently using the plebian form of her maiden name at the time.)

To cover his bewilderment, Fulton immediately began talking about his own projects. The Marquis, thinking him mad, cut the interview short. At this period, a few people meeting Fulton for the first time did think him crazy. He would launch on almost total strangers a tumult of his ideas and plans. Yet, such unsought confidences resulted from loneliness, not madness.

Fulton's usual practical view of life extended to the opposite sex. In Philadelphia, the daughter [4] of John Ross had attracted him. There is mention of a rector's daughter whom he had met in London, as well as of a nameless brunette in Manchester. Yet, he had evidently seldom even considered marriage. Poverty and his peripatetic personal life had offered him little opportunity for women.

Fortunately, Robert Fulton soon found a cure for his loneliness. He had a letter of introduction from Benjamin West to an American in Paris named Joel Barlow. The inventor moved into the hotel where Barlow and his wife Ruth lived. The three hit it off immediately. The childless Barlows were a decade older than Fulton and soon treated him as a favorite younger brother or even son. In Barlow the younger man found the father he had never really had. In Ruth Barlow, he found an older sister, aunt, and female friend who suited admirably his psychological needs at this time. A remarkable three-way friendship began in 1797 that lasted for the rest of their lives.

Joel Barlow was already everything that Robert Fulton wanted to become—wealthy, cultivated, respected, a man of affairs. He laid the groundwork for his modest fortune by running blockades of France from 1792 on. He rationalized his opportunism by arguing, as James Woodress, his latest biographer, puts it, that "by helping to keep the supplies flowing to France, Barlow was striking a blow for freedom. If an American businessman could aid these worthwhile objectives and also make a profit, why should he not do it?" [5]

Barlow had many advantages. He was the only American who also held French citizenship as a result of aid to the rebels during the French Revolution. He had both open and clandestine connections with many English. In at least one blockade-running episode, he had actually leased an English ship, *Cumberland,* to do the job. Throughout the furious comings and goings of rival factions during and after the French Revolution, Barlow had also managed to keep his lines of communication open. In short, he was ideally situated to be a man of business during such hectic times.

In later years, this period of his life aroused considerable speculation and curiosity, and all his transactions have never been unearthed. However, Woodress reports on enough of them to indicate how he operated. His commission as an agent amounted to eight to ten percent of the value of the cargo. He also owned some cargoes outright or owned part of them, and made much money this way—although he also risked large sums if the British should capture any of his ships.

He often worked as an agent for James Swan, the chief American merchant in Paris (called a "corrupt, unprincipled rascal" by James Monroe) and received incoming goods from American and Scandinavian merchants. By the time Barlow settled in Paris more or less permanently in 1795, he had grown comparatively rich.

With an estimated $120,000 in safe investments, Barlow became a statesman and philosopher. He was considered the most cosmopolitan American of his age. As a poet,[6] he enjoyed an immense reputation in the United States, although few of his verses remain readable today. As a savant, he published letters on mechanics, education, scientific agriculture, and many other subjects.

Furthermore, his youth and early manhood paralleled Fulton's. Both were born to poverty on American farms. Both had lived by their wits during their early days. Both had somewhat elastic scruples.

When younger, Barlow had been involved in a financial scandal. As European agent of the Scioto Association, he had sold dubious deeds to land in what is now Ohio to gullible foreigners, yet he escaped with his reputation still clean even after an investigation. From then on, he kept his financial affairs quiet. As we have seen, so did Fulton.

Both men had cultural interests—Barlow in literature, especially poetry; Fulton in art. Barlow also soon developed a Pygmalion-like attitude toward Fulton; teaching him, guiding him, shaping him. The younger man responded eagerly because he well knew that his limited formal education handicapped him seriously in pursuing a career.

Not a hint has surfaced of homosexual relations between the two. Barlow longed for a son. Fulton needed a father. These unusual men played the roles to each other's complete satisfaction.

With Ruth Barlow, Fulton served as escort during her husband's frequent absences, sympathetic supporter during her many illnesses, and lively and entertaining companion. Again, no evidence exists of sexual relations between them. Although Ruth Barlow would disapprove strongly when Fulton eventually married, her displeasure was more like that of an aunt or older sister who considered no woman good enough for the beloved nephew or brother.

The trio would live together for seven years; in 1797, the Barlows and Fulton had each planned to return to America momentarily. A big reason why they all delayed the step for years was that they found each other's company so congenial in Paris. By 1800, they had all grown tired of lodgings, so Barlow decided to buy a house. His choice was 50 Rue de Vaugirard, across from the Luxembourg Gardens. At the age of 35, Fulton moved with the Barlows into the first house he had lived in with any permanence since his teenage days in Lancaster.

He must have been overwhelmed when he saw it. First, a long frontage of one hundred and forty feet confronted him on the street, along which rose a solid building, occupied by kitchen, servants' quarters, and stables. A splendid gate gave him a hint of what lay ahead. Through the gate, he entered a courtyard with upper and lower levels connected by a handsome terrace. To his right and left were wings running from the street-front building toward the main building. The house itself extended the entire width of the property, to make a hollow square with the wings and street structures.

Fulton entered the mansion through an elegant vestibule that gave upon a salon seventeen feet high. On each side of the main entrance he explored complete living quarters. He would occupy one, the Barlows the other. Behind the salon was a grand dining room and a large library.

The full establishment contained twenty-two sleeping rooms, stables for twelve horses and could shelter five carriages. The twenty cellars under the house could hold five hundred wine casks. Behind the house, extensive gardens covered an acre.

A nobleman had built the place only a decade before at an expense of about $100,000 (not counting the land). The revolution had begun almost as soon as he had finished it, and he had to flee. He had returned to Paris, but was too poor to live in his distin-

guished *hôtel*. Not many other people could afford it either, and
he had trouble selling it. Businessman Barlow bought it as an investment for only a quarter of its original cost.

Although the Barlows and Fulton never did use or furnish
more than a small portion of this establishment, Robert particularly reveled in it. His artist's eye appreciated the architecture. Its
elegance and solidity satisfied psychological needs. It would serve
him as a model for his style of living during the rest of his days.

Barlow did furnish one room—the library—so completely that
when he brought his books back to America, he had one of the
best collections in the United States, public or private.

That autumn and winter of 1800, he and Fulton settled down
to a course of study that was, in effect, the completion of the
younger man's college education. Barlow tutored him in French,
Italian and German literature and philosophy. Together they
studied higher mathematics and immersed themselves in physics
and chemistry. Barlow taught with both intuitive and professional
skill and the father-and-son relationship between the two spurred
his intuitive drives. After his graduation from Yale, in 1778, Barlow had taught school briefly. (During his checkered career he
had also edited a newspaper, practiced law, engaged in politics
and diplomacy besides his business affairs, and had even been a
chaplain in the American Revolutionary Army.)

At about this time, Fulton painted a portrait of Barlow that may
be the best art work he ever did. The picture shows a handsome,
self-assured man with a Roman nose, large, well-shaped mouth,
well-defined brows—in short a person not unlike Fulton, but
eleven years older.

The two men also spent the fall and winter discussing and improving the conception of a project that had become Fulton's
major preoccupation in France, displacing his concern with
canals. As early as December 13, 1797, he wrote the Directory, in-
forming it, "having in view the great importance of lessening the
power of the English fleet, that he had a project for construction
of a mechanical *Nautilus*."

The *Nautilus* was a submarine, suspiciously similar in conception to one developed earlier by a fellow American, David Bushnell. Although Fulton did not claim complete originality for his
canal ideas (nor would he for the steamboat), he never acknowledged any indebtedness to Bushnell. A major explanation of this

lack may lie in the fact that Bushnell never forced him to. For reasons that remain obscure, the originator of the submarine chose to migrate from Connecticut to Georgia after the Revolution and live incognito under the name of Bush. His will revealed his true identity after his death.

One reason for such action may be the criticism that arose over the very concept of the submarine. Many in that day considered it a barbarous method of war, akin to the way contemporaries now look upon poison gas or the atomic bomb. Indeed, Fulton encountered this same attitude toward it from the ruling French Directory, which included some of the people who had unleashed the Reign of Terror.

Yet, Bushnell's technical achievement with the submarine remains almost miraculous.[7] Nothing in his background prepares us for his discovery and application of basic principles that still underlie the submarine today.

It's possible, but not probable because of the poor libraries and research resources in the American colonies, that Bushnell knew of earlier experiments with submarines in Europe. In the sixteenth century an Englishman named William Bourne, a gunner under Admiral Sir William Monson, had demonstrated how a craft made with a wooden frame covered with leather could be submerged and rowed beneath the surface; water was used as ballast and the volume of the vessel was changed by contracting the sides with hand vises.

A Dutchman, Cornelius van Drebbel, followed up on Bourne's research. He designed an underwater ship with a leather bulkhead; when the bulkhead was withdrawn, water poured in through holes in the ship's sides until the upper deck was covered. Then, with oars paddling, the vessel moved along partially submerged. When the leather bulkhead was screwed back, the ship became totally buoyant again—hopefully. Legend has it that King James I traveled from Westminster to Greenwich in one of Drebbel's wood-and-greased-leather ships. Drebbel also installed air pipes with a purifying system worked by bellows.

After Drebbel there came a series of designs and, by 1727, fourteen had been patented in England. But the most ambitious and warlike was designed by two French priests, Fathers Mersenne and Fournier, of the Order of Minimes, who devised a vessel with wheels to run along the sea bed, air pumps, phosphorescent lighting, big guns and an escape hatch. After that came experiments with leather bottles filled with water ballast which could be expelled by hand.

In 1773 a carpenter named Day experimented with stone ballast, the stones being attached to the outside of the hull and released from inside. Day dived successfully in Plymouth Sound, England, before moving to deeper water. With extra ballast he sank into twenty-two fathoms and never reappeared.

David Bushnell was born in 1740 on his father's farm in what is now Westbrook, Connecticut. He is said to have been studious and reclusive. On the death of his father, when David was twenty-eight, he sold his inheritance and moved to Saybrook to prepare for college. The Yale catalog of 1771 shows him to have enrolled in the college. He was thirty-one. He received his B.A. degree on July 25, 1775 at the age of thirty-five.

He is principally remembered for three actions during the Revolution: a submarine attack on the British in New York Harbor, an attack on the frigate *Cerberus* off Black Point in Long Island Sound in August, 1777, and the Battle of the Kegs in the Delaware River at Philadelphia at Christmas, 1778. None was conclusively successful, but the last caused much confusion among the British.

Writing in 1787 to Thomas Jefferson, at Jefferson's request, Bushnell recalled:

> The first experiment I made, was with about two ounces of gun powder, which I exploded 4 feet under water, to prove to some of the first personnages in Connecticut that powder would take fire under water. The second experiment was made with two pounds of powder, inclosed in a wooden bottle, and fixed under a hogshead, with a two-inch oak plank between the hogshead and the powder; the hogshead was loaded with stones as deep as it could swim; a wooden pipe descending through the lower head of the hogshead, and through the plank, into the powder contained in the bottle, was primed with powder.
>
> A match put to the priming, exploded the powder, which produced a very great effect, rending the plank into pieces; demolishing the hogshead; and casting the stones and the ruins of the hogshead with a body of water, many feet into the air, to the astonishment of the spectators. This experiment was likewise made for the satisfaction of the gentlemen above mentioned.[8]

In 1771, Bushnell as a freshman at Yale came upon the first of
the two basic submarine principles that Fulton later adopted as his
own. Bushnell had doubted an instructor's statement that gunpowder would not explode under water. He disproved him in a
way that seems obvious, but was unusual then, by actual experiment. What's more, he discovered that the same amount of gunpowder will do more damage under water than above—that's because the detonation force does not dissipate as easily in water as
in compressible air.

With the Revolutionary War approaching, Bushnell tackled the
problem of how to attach gunpowder (or bombs) to the keels of
British warships. He needed a delivery system that could do the
job without detection—in short, some sort of boat that would
move under water. Earlier, as we have seen, others had dreamed
of submarines, but no one had been able to build a practical vessel. Bushnell came closer than any predecessor. He constructed it
secretly on the Connecticut River in 1775. Although it lacked a
periscope and electric or atomic power, it had many of the basic
principles of today's submarine. What's more, he used the screw
propeller, the first known practical application for any vessel either under or on the surface of the water.

Bushnell called his craft the *American Turtle* because it looked
like one. It was six feet high and held only one man. Thin and
tall, it also resembled a large pocket flask standing on end. Bushnell said it appeared like the shells of two tortoises pressed
together. A little conning tower rose where the heads would be.
When the vessel floated, you could see only the conning tower
which looked like a large brass hat. As the operator opened a
valve to let water into a tank on the bottom, the added weight
pulled the ship fully below the surface. Pumps run by foot power
could force the water out again when the operator wanted to rise.
Likewise at his feet was a treadle similar to one used on a nonelec-
tric sewing machine which moved the propeller at the rear. The
vessel had other propellers to move the craft sideways or up and
down.

Ventilators rose from the conning tower, one to carry out bad
air from the top of the boat and the other to suck good air into
the bottom. Valves shut automatically when the *Turtle* submerged.
Other automatic valves closed apertures if the windows in the
conning tower broke. Bushnell even used the modern theory of

redundancy so common on space vehicles. In each case, a second set of valves could be operated by hand if the automatic units failed.

A perforated plate over every valve strained the water to avoid clogging any moving part. All holes through which apparatus extended were made of brass pipes into which iron rods fitted exactly, a marvel in those days of crude iron work. Oil in the joints prevented rust or leakage. Bushnell even invented a barometer to show depth under the surface. It was a sealed tube open at one end to the water, on which the pressure was recorded by a floating indicator.

And how did the operator find all his controls and other apparatus he had to manipulate in the dark when submerged?

> Particular attention was given to bring every part . . . both within and [extending] without the vessel before the operator, and as conveniently as could be devised; so that everything might be found in the dark, except the water gauge and compass, which were visible by the light of the phosphorous, and nothing required the operator to turn to the right hand or to the left, to perform anything necessary.[9]

Attached to the rear of the *Turtle* was a bomb, with a rope from it to a spike that projected through the conning tower. Bushnell planned to operate his vessel at night, coming awash to the enemy ship, with only the brass hat of the tower visible. When it reached the enemy vessel, the *Turtle* would submerge and sneak to the keel. Pounding from inside, the operator would hammer the spike into the wooden ship. Next, he would release the bomb that would float against the enemy's side and stay there, secured by the rope leading to the spike. A clock, which had been set in motion automatically when the bomb was released, ticked menacingly as the operator pumped furiously to turn the screw and reach safety before the bomb exploded.

Bushnell persuaded America's leaders—Franklin, Washington, Jefferson, Connecticut's Governor John Trumbull and many other patriots—to provide him money, troops and supplies. In 1775, America had no navy, certainly nothing that could remotely match Britain's overwhelming seapower. The submarine and bomb might be the answer.

As we have seen, they did not do the job, but America's "secret weapon" frightened the British, and its existence was widely known among leading Americans, and almost certainly to Joel Barlow. Barlow's freshman year at Yale coincided with Bushnell's senior year. With a total student body of less than 150 at the college, Barlow must have known the older man who had become a local celebrity. In 1778, the year of Barlow's graduation, Bushnell submitted the detailed description of his experiments, his apparatus, and his attempts against the British fleet as the thesis for his Master's Degree. It is reasonable to conjecture that Barlow had access either to this thesis or to a similar document that Bushnell had sent to Jefferson at Jefferson's request when he was ambassador to France. Barlow knew Jefferson intimately. Through Barlow, Fulton must have learned most of the details of the *Turtle,* although neither ever admitted it; the subject probably came up frequently at the dinner table or fireside.

Some biographers of Fulton have speculated that he got the plans directly from Bushnell himself. This is not likely. As far as is known, the older man never left American shores. No records show that the two had ever met before Fulton had sailed for England in 1787, and his attention was then focused on painting, not invention.

By the 1790's, Bushnell had disappeared. He had become a physician named Bush who lived in Georgia and was to grow wealthy in land speculation.

Fulton took over the submarine. He had written tracts arguing that free trade would abolish war. He sent these to the French rulers who ignored them. Their refusal to accept this truth convinced him, he wrote, that "society must pass through ages of progressive improvement before the freedom of the seas could be established by an agreement of nations that it was for the true interest of the whole." Believing that navies caused the world's war troubles, "I turned my whole attention to find out means of destroying such engines of oppression by some method which would put it out of the power of any nation to maintain such a system, and would compel every government to adopt the simple principles of education, industry, and free circulation of its produce." Needed were two things to reach this objective, and he blandly described them: "First, to navigate under water, which I soon discovered was within the limits of physics; second, to find an easy mode of destroying a ship, which after some time I discovered might be done by the explosion of some pounds of powder under her bottom." [10]

Although professing idealistic motives in proposing the submarine in 1797, Fulton asked for rewards that could have made him a rich man if the French had granted them and if his design had worked. He had organized a stock company to build the submarines at its expense. All the government had to do was to pay the company 400 livres per gun (about $15,000 in today's equivalent) for each British warship over forty guns that they sank, allow them to keep all prizes captured, and give them an exclusive right to such boats, unless the French decided to build them themselves. Then the government would pay the company 100,000 livres ($3.6 million) for each one built.

"As a citizen of the United States," he continued, "I hope that it may be stipulated that this invention or any other similar invention be not employed by the government of France against the United States, unless the American government first use it against France." [11] He also asked that the Directory give his company's sailors commissions stating that if they were executed or treated other than as ordinary prisoners, the French would retaliate on British captives in their hands. Fulton added this last provision because he suspected that a submarine would be considered against the rules of war.

On February 5, 1798, the Directory turned down Fulton's proposal because it refused to give his sailors commissions. "The government," the Minister of Marine explained, "cannot publicly avow men who undertake this type of operations . . . and this would be in some sort to scratch out from the code of war the just punishments inflicted on those who are naturally inclined to fight in an atrocious manner." [12]

Fulton impatiently awaited the appointment of a less squeamish minister, correctly foreseeing that Napoleon would soon rocket to power. While waiting, Fulton wrote to Napoleon, calling attention to the civil advantages of his small canal system. He also included his tract on freedom of trade, not neglecting to lecture the most important man in France on political reasoning, with the conventional liberal thinking of that time derived from the writings of Adam Smith. Fulton saw nothing incongruous in writing to Napoleon in this vein:

> Among all the causes of wars, Citizen General, it is true that we see every day disappear those that relate to kings, priests, and the things which accompany them. But, nevertheless, republics themselves are not exempt from melancholy quarrels, inasmuch as they

do not separate themselves from the erroneous system of exclusive commerce and distant possessions. All who love their fellow-men should attempt and seek to destroy these errors.[13]

He urged Napoleon to use his prestige to "favor projects the execution of which would render happy millions of men . . . If success crowns the efforts of France against England, it will only remain for her to terminate this long war gloriously by granting freedoms to trade . . ." Napoleon never replied.

Where Bushnell lacked persistence, Fulton had it in full measure. Rebuffed the first time by the Directory, Robert waited until a new Minister of Marine assumed office. After that had happened, he tried again on July 23, 1798, concluding his offer by pointing out that the destruction of the English navy would assure the freedom of the seas, and the nation which had the most natural resources, France, would alone hold the balance of power in Europe. The Minister convened a board of technical men to whom Fulton submitted his plans for the *Nautilus*.

According to this proposal, the boat would have the shape of an imperfect ellipsoid, with an overall length of twelve feet, three inches and extreme beam of six feet, four inches. Beneath the ellipsoid was a hollow iron keel one foot, eight inches in height, running to within thirty-nine inches from the bow. The keel contained the ballast so that the difference between the weight of the flotation and that of the water displaced by it should be only about nine to twelve pounds. The only communication with the interior of the keel lay in the two parts of a suction and force pump which, by means of a hand crank, would permit the introduction or removal of water. Because the excess in buoyancy of the *Nautilus* was small, the introduction of only a small volume of water would make it sink. Conversely, the expulsion of only a small quantity would make it return to the surface. On the forward and top part of the *Nautilus* perched a spherical dome pierced with port holes covered by thick glass for observation and a manhole by which the crew could enter or leave.

For propulsion, Fulton proposed a screw as Bushnell had also done. The screw was at the stern and directly ahead of the rudder, operated by a hand crank and gearing that turned a shaft passing through a stuffing box. A man pumped on a treadle-like arrangement to turn the crank that moved the screw. Plunging was accomplished by pumping water into the keel, while submer-

sion at a given depth, if the boat was in motion, was to be attempted by means of two inclined planes attached to the sides of
the steering rudder. The angle of these planes could be altered
from within, giving an upward or downward direction to the boat.

Fulton obtained motion on the surface by a fan-shaped sail
which, with the supporting mast, could be folded down to the
deck and then, before submersion, covered with envelopes like
the wings of a fly. Fulton estimated that he could work the boat
with a crew of three, of whom one would be himself.

Fulton displayed a model for the commission of distinguished
scientists established to study the invention. Their report of September 5, 1798 describes the device which is essentially an enlarged *American Turtle*. Most of the faults that the commission
found with the conception arise where Fulton had departed from
Bushnell's ideas. They objected to the horizontal rudders, so Fulton remedied that problem with Bushnell's vertical propeller.
They objected to his method of renewing the air, so Fulton
adopted Bushnell's solution. And Fulton added Bushnell's barometer when the commission complained that the crew had no way
to know their depth. Yet the commission was impressed. "The
weapon conceived by Citizen Fulton is a terrible method of destruction, since it acts in silence and in a manner almost inevita-
ble," they wrote. "The weapon is undoubtedly imperfect; it is the
first conception of a man of genius." [14] They recommended that
Fulton get the money to build a full-sized submarine.

Still, the Directory did not act. In October, Fulton tried to speed
matters by sending one of the Directory members, Barras, an attack on the "monstrous government of England." His scheme, he
argued, only seemed revolting because it was extraordinary.

Apparently deciding that he needed more than one arrow in his
quiver, Fulton allowed Josiah Gilpin to carry news of his secret
weapon to London, to Stanhope, Pitt's brother-in-law, and others.
He did this while assuring the French of his undying hatred of
the English government. At the same time, he told Gilpin that he
was no French partisan, his only desire being "to promote the interests of mankind . . . I cannot unite with any party or polity,
nor will I aid them unless I clearly see that an obstacle between society and a lasting peace or improvement can be removed." [15]
Hearsay also hints that he tried to build interest in the submarine
in Russia, whose political views he claimed to abhor.

7 DOWN IN THE DEPTHS WITH THE *NAUTILUS*

While Fulton waited for almost any government to help him finance the submarine, he cast about to find ways to raise money without any government's help. He tried to get orders to paint portraits, but Barlow alone gave him one, more a tactful form of charity than a true commission.

He also took out several patents, among them the "import" of Cartwright's rope-making machine, previously mentioned. During the Napoleonic troubles, France for a time did not recognize English patents. It was legally, if not ethically, permissible to take out French patents on English inventions. Fulton had done so on the rope-maker. What he next did, however, was even questionable legally. He sold the French patent to a fellow American, Nathaniel Cutting. This action would embarrass Fulton years later in America.

Finally, he hit upon a real money-maker. On April 26, 1799, he took out a French import on an English patent held by Robert Barker for a panorama. Barker, an Edinburgh portrait artist, painted huge pictures on domes and charged admission. His Panorama of Edinburgh was exhibited in the Haymarket in 1789. He next painted a view of London which was shown in 1792. Finally, in 1793, he took a lease on land in Leicester Square and erected three panoramas, the largest being ninety feet in diameter. Fulton must have seen all or some of these.

By December 8, 1799, he sold a share in his French patent rights to James W. Thayer, a compatriot. They secured a plot of land on the south side of the Boulevard Montmarte, erected a building forty-six feet in diameter to house the panorama.[1] He

painted the "Burning of Moscow," and opened it to the public
early in 1800. This, of course, did not prophecy the fire oc-
casioned by Napoleon's invasion in 1812; it represented one of
several earlier conflagrations that had devastated the city. Although Fulton had disposed of all his interests much earlier, it
was so successful that it ran until 1814 and was even commemo-
rated in a Parisian street ballad, one stanza of which goes:

> Paris more than any place
> Rejoices in a legitimate success.
> A clever man will show it you
> For one franc fifty centimes.
> And everybody goes or is going
> To the pano, pano, panorama.[2]

Fulton at last had some financial breathing room. But still the
Directory did nothing. In despair, he tried to interest Bavaria in
the submarine, through the auspices of the ubiquitous Herr Schemelpeninck who had secured the freedom of "Madame François."
That, too, came to nothing. Next, he went to Holland, where a
M. de Vanstaphast offered to back him, but the government gave
him no encouragement.

He left that country, apparently with relief, when word reached
him that yet another overturn in the French government on November 9 and 10, 1799 had put Napoleon in as the First Consul
and virtual dictator of the nation. Fulton called quickly on the
new Minister of Marine, none other than P.A.L. Forfait, a naval
architect, who had served on the first commission that had reported favorably on the submarine.

Forfait replied ambiguously on April 15, 1800 to Fulton's new
request:

> It cannot be disguised that the *Nautilus* is a machine not yet in use
> and that it infringes in several points the laws of war. It would be
> dangerous, especially at this moment when so great a number of
> Frenchmen are in the power of the English, to express any kind of
> menace in the Commission. In granting it pure and simple, that is
> to say, in acknowledging as combatants the men serving on the
> Nautilus and the Nautilus itself, I think that that ought not to
> create more fear than the menace of reprisals can give security for.[3]

Forfait may have given money to Fulton under the table because an April 10 letter from the inventor indicates that he was

having the vessel completed in the workshop of C. Perier in Rouen. Yet, if Forfait did provide funds, they were not enough. Fulton and probably Barlow also helped finance the boat. By July, it was nearly finished, and launched on July 24. Fulton went aboard himself with two helpers. The first plunge lasted eight minutes and the second seventeen minutes. The trials so encouraged him that he dashed off this letter to Forfait:

> Yesterday (July 29, 1800) I tryed my experiment with the Nautilus in water 25 feet deep and have succeeded to Render the sinking and Rising easy and famelior, the Current which was at least one League per hour together with the want of suffecient experience, prevented me making the movements under water which I desired, however time will perfect that part of the operation, having succeeded to sail like a common boat and plunge under water when I think proper to avoid an enimy   it may be suffecient at present to render an operation against the enimy successful, this day I propose to set off to Havre and hop to arrive there on the fourth, be so good as to send me An order for the powder I may want which will be from 8 to 10 Quintals.
>
> I have not heared any thing of the letter of protection from the Premier Consul be so good as to spech to him on that subject and let me know his determination.[4]

At Le Havre, Fulton made more trials in August, 1800. Ruth Barlow accompanied him to the port because she had been ordered there by her physician for the sea bathing and air. Her husband's letters to her reveal that Fulton himself was financing part of the expense:

> Tell Toot [the Barlows' pet name for Fulton] he shall have the $1000 in a day or two, but Thayer has not paid according to his promise. The pictures [the panorama] go not well—50 or 60 livres a day for both—and at this season! But the excessive heat prevents everybody from stirring out, especially on the Boulevards, and in the daytime.[5]

The danger of the experiments also alarmed Barlow. To his wife he wrote in terms remarkable for that reticent era: "Always repeat to him how much I love him; you cannot tell him too much of it." Barlow asked Ruth to remind Fulton that his body was more valuable than any machine he could devise "and that unless he could create me one in the image of himself, he had better preserve his own automaton. Read this lecture to him, or a better one on the preservation of his health and vigor, every morning at breakfast." [6]

Barlow was also capable of joshing about Fulton. On another occasion when Fulton had accompanied Ruth on a futile health trip to the baths, Barlow urged his wife to "give the waters fair play, trous des Capucins & all." The "trous des Capucins" refers to a religious order who sought alms at the baths and had a remarkable reputation for enabling the women partakers of the waters to become pregnant. Barlow continued: "I reckon those ladies who come for the purpose of making babies may find the same instruction & spiritual edification from Toot as they used to do from the barefooted brotherhood."

Fulton next decided to try his vessel on the high seas and proposed to go to Cherbourg for that purpose. Alas, more bureaucratic delays made it impossible for him to go quickly because he had no commission or passport. He asked Barlow for help, and the older man replied on September 6, 1800:

> Dear Fulton—Your letter of the 16th came yesterday about 4 o'clock, too late to see the Minister, and this morning he seems to have got up wrong end foremost. I went to his porter's lodge at 9 o'clock and sent up a letter concise and clear, explaining the affair and telling him I should wait there for an answer, or for leave to speak to him . . . I always doubted whether this Government would suffer your expedition to go into effect. It is possible they have reserved to themselves this method to prevent it, always in hopes before that your preparatory experiments would fail, or that your friends and patience would be exhausted.[7]

However, Barlow did finally get the desired commission from reluctant Ministry officials and commented:

> Your old idea that these fellows are to be considered parts of the machine, and that you must have as much patience with them as with a piece of wood or brass, is an excellent maxim . . . I will take care that it shall not be forgotten by the writer of your life, who, I hope, is not born yet.[8]

On September 12, 1800, Fulton set sail in his tiny submarine to sink the world's most powerful navy. Many people thought him as mad as Don Quixote, although few did who had met and talked with him. Handsome, voluble, and persuasive, he could make most listeners think he could accomplish anything. They even accepted the incongruity that Fulton had qualms that his craft could sail like an ordinary boat—while at the same time he had no doubt that it could blow up British ships.

To the inventor's relief, the *Nautilus* did sail adequately in excellent weather to a harbor seven and a half miles from the Marcou Islands. Two British brigs had been blockading there for months. Frustrating storms delayed his ultimate trial. Finally, relative calm permitted the attempt. He and his two helpers sailed out at night as close as they dared to one of the brigs. They furled the sails and lowered the mast while the cockle-shell pitched and tossed in the swell.

All climbed below. After securing the porthole, Fulton pushed a pedal to allow water into the keel's tank. The *Nautilus* sank like a stone. The three men huddled in a tubular room with a rounded iron roof about four feet high—like the inside of a boiler. Seizing handles that turned the propeller, the two sailors struggled, but the vessel moved only with difficulty, despite the fact that the tide ran out toward the British ships. Before they had come close to the target, the tide had turned. Cranking was now useless. Fulton cast anchor, raised the ventilator pipes, and announced that they would wait under water for six hours until the tide turned again.

As the minutes crawled by, Fulton must have blown out the candle, both to save precious oxygen and to avoid staring at his crew. They were all thinking the same thing: Will the air last? Will the *Nautilus* be able to rise to the surface? The two helpers must have been wondering why they had ever allowed this crazy man to talk them into such a hare-brained venture.

From time to time, Fulton would light the candle to see his watch, then hastily blow it out. His bleak companions and the moisture condensing inside the tank could not have reassured him.

At last the tide turned. The sailors labored again at the propeller cranks, and the *Nautilus* inched forward. When Fulton judged they had neared the brig, he risked raising her to the surface for a peek. What a relief! She did rise. Then disappointment supplanted the relief. The submariners could see nothing in the half light of dawn. The British brigs had fled. Fulton ordered the mast raised and the sails unfurled to return to shore.

On at least one other occasion, he tried again, with the same results. The British secret service had watched his movements, undoubtedly alerted by Fulton's messages to Stanhope. An Admiralty order warned the Havre squadron of "Mr. Fulton's plan for destroying ships." H.M.S. *L'Oiseau's* captain had answered, "I shall be very much on my guard." [9]

Although Fulton had failed to sink anything, not even a canoe, he claimed a great success for his submarine because he had proved that underwater navigation was possible. He even reminded the Barlows' friend, Tom Paine, that he would not require a whale to move his vessel, as the radical firebrand had jested in his heavily satirical manner.

Fulton's friends, the commissioners, recommended that the government grant him 60,000 livres (nearly $22 million in today's values, which he never got). They won him an audience with Napoleon. The temptation is great to imagine a colorful scene with the dictator. But the fact is that the meeting was brief, too short for Fulton to work the magic of his charm and persuasion. Bonaparte had probably already developed an immunity to that kind of approach, in any event. Some biographers and several French naval historians have criticized Napoleon's lack of enthusiasm for the "means to achieve world dominion." The First Consul was so little impressed that he even later forgot the interview, as we shall see. Actually, as Fulton had no reliable means for propelling the submarine, and had failed to prove the military effectiveness of his invention. Napoleon was undoubtedly correct in his decision.

In a report to the commissioners appointed by Napoleon to appraise the invention—Gaspard Monge, Comte de Peluze, mathematician; and Pierre Simon, Marquis de Laplace, mathematician and astronomer—Fulton's own version of his experiments contrasts with reality:

ROBERT FULTON

To *Citizens Monge & LaPlace, members of the National Institute*

Citizens,—Not having had the time to busy myself with the drawings and description of the latest changes that I have thought fit to make in my Nautilus, I take the liberty to recommend the model of it to your examination as the best means of enabling you to judge of its form and combinations.

Although having exact details of experiments, I shall limit myself to rendering here a succinct account of the most important of them.

First experiment.—The Nautilus is 20 feet long and 5 in diameter and according to the calculations of C^{en} Guyton it will contain a quantity of air sufficient for 3 men and a candle for three hours.

Second experiment.—On the 6 Fructidor (24 Aug. 1800) I plunged in the basin at Havre to the depth of 15 feet having with me two people and a lighted candle; we remained below the surface for the space of one hour without experiencing the slightest inconvenience.

Third experiment.—On the 7th (25 Aug.) I tried to manoeuvre the Nautilus by means of wings 4 feet diameter like the sails of a wind-mill; to this end at first I placed on the bridge two men with oars; they took 7 minutes to row about 90 toises (192 yards), the length of the basin; then I ordered the same 2 men to set the sails and in 4 minutes the Nautilus covered the 90 toises to the starting place;— I proved by this that the speed of sails to that of oars is about 2 to 1 and that these sails are very suitable to manoeuvre a boat under water. The success of this experiment has given me several new ideas which I hope will facilitate much the use of carcasses of powder or torpedoes.

Fourth experiment.—On the 8th (26 Aug.) I tried balancing the Nautilus under water in such a way as to prevent it rising towards the surface or descending to the bottom, meanwhile advancing. This is executed by means of a pair of wings placed horizontally on the front of the Nautilus and which communicate with the interior. By turning these wings from left to right the Nautilus is made to descend below the water, in turning them from right to left, it is raised to the surface. My first trial was unfortunate, in not having placed the boat in the necessary trim in order that the wings could act. The next day I had a decided success and I kept my Nautilus below water at a depth of about 5 feet whilst it covered a distance of 90 toises, about from one end of the basin to the other. This day I made several movements under water and I observed that the Compass acts as well under water as at the surface. The three people who have been my companions during these experiments are so familiarized with the Nautilus and have so much confidence at present in the movements of this machine that they undertake without the least concern these aquatic excursions.

Having thus assured myself of the ease of emersion and of submersion of the Nautilus and all its movements as well as the effect on the compass, on the 9th (27 Aug.) I *half filled* an ordinary barrel and placed it at anchor in the harbour at about 200 toises (426 yards) from the jetty;—I seated myself then in an ordinary boat at the distance of about 80 toises and placed in the sea a torpedo containing about 30 lb. of powder; the torpedo was attached to a small rope of 100 toises; the current going under the barrel, the torpedo passed without touching it; but turning the helm of the boat in which I sat, I made it go obliquely till I saw the torpedo exactly under the barrel; I then drew back the cable till at last the torpedo touched the barrel; at that instant the battery went off, the powder exploded and the barrel was reduced to fragments being lost in a column of water 10 feet in diameter that the explosion threw into the air to the height of 60 or 80 feet.

On the 25 of the same month (12 Sept.) I left Havre for La Hogue and in this little voyage, my Nautilus sometimes did a league and a half (4½ miles) per hour, and I had the pleasure of seeing it ride the waves like an ordinary boat.

On the 28th (15 Sept.) I put into a little harbour called Growan near Isigny at 3 leagues from the islands of Marcou. On the 29th the equinoctial gales commenced and lasted 25 days. During the time I tried twice to approach two English brigs which were anchored near one of the islands, but both times, whether by accident or design, they set sail and were quickly at a distance. During one of these trials I remained during the whole of one tide of 6 hours absolutely under water, having for the purpose of taking air only a little tube which could not be perceived at a distance of 200 toises.

The weather being bad, I remained 35 days at Growan and seeing that no English vessle returned, and that winter approached, besides my Nautilus not being constructed to resist bad weather, I resolved to return to Paris and place under the eyes of Government the result of my experiments.

In the course of these experiments there has come to me a crowd of ideas infinitely more simple than the means that I have employed hitherto and in an enterprise so new and without precedent one ought to expect that new ideas should present themselves, tending to simplify the execution of the great object in view.

As to myself I look upon the most difficult part of the work as done. Navigation under water is an operation whose possibility is proved, and it can be said that a new series of ideas have just been born as to the means for preventing naval wars or rather of hindering them in the future; it is a germ which only demands for its development the encouragement and support of all friends of science, of justice and of society.

Health and respect.
ROBERT FULTON [10]

The First Consul merely ordered the Marine Ministry to study his proposal. The optimistic and sanguine Fulton at first read into the tepid commitment assurances of success at last. By December, 1800, however, his high hopes had turned sour, and he wrote the Marine Minister a rare, intemperate letter:

> You will permit me to observe that although I have the highest respect for you and the other members of the Government, and although I retain the most ardent desire to see the English Government beaten, nevertheless the cold and discouraging manner with which all my exertions have been treated during the past three years will compel me to abandon the enterprise in France if I am not received in a more friendly and liberal manner.[11]

This is the only letter in French found in the government archives wholly in Fulton's handwriting. Others were copied by someone and signed by the inventor. This exception indicates he wrote it in white heat without the counsel of his wise and diplomatic friends, Barlow, Monge, and LaPlace.

At about this time he also threatened the French government. "I sincerely hope for the honor of France," he wrote, "that I will not meet the objections of narrow spirits or little intrigues which will put me to the necessity of publishing the principles of the Nautilus and their happy consequences, or to seek in Holland or in America the encouragement which I hoped to find in France and which liberty and philosophy demand." [12] (Note the disingenuous reference to Holland. As we have seen, the Dutch government had already spurned him.)

Perhaps he made this threat to counteract the possibility that the French had seen a published account of Bushnell's submarine. The *Transactions of the American Philosophical Society* had published in 1799 his full and exact description. Because this journal sometimes circulated among French scientists, Fulton had reason to fear that they would see it. Did he back off on his claims for originality? Not at all. He went on as though nothing had happened.

Forfait vacillated, but by March 28, 1801, he announced that the government would credit Fulton's account with 10,000 francs (about $3.5 million today), far less than the 60,000 livres recommended originally. The Forfait letter explains the terms: [13]

THE MINISTER OF MARINES AND OF THE COLONIES
To *Monsieur Robert Fulton*
  Rue de Vaugirard, No. 50, Paris

I announced to you, Sir, on the 8th Ventose that the First Consul had authorised me to accept your proposition relative to the *Nautilus*. You will have seen by that letter that you will in consequence be credited with the sum of 10,000 francs to repair this machine, construct the auxiliaries, and to convey at your own expense, the *Nautilus* to Brest.

It has been decreed that you will be allowed for the destruction of the Enemy's vessels, according to their armament, as follows:

400,000 francs for those of more than 30 guns.
200,000   "   "   "   "   "   " 20 " up to 30 guns.
150,000   "   "   "   " from 12 to 20 guns.
  60,000   "   "   "   " 10 guns.

This power is the minimum below which you will have no power to return claim.

By your letter of the 12th Ventose you declare your acceptance of these conditions and I give the order to put to your account the sum of 10,000 francs by means of which you must put in order the armaments, the equipment and the dispatch of the *Nautilus*.

There exist several means of determining in an authentic manner the destruction of the enemy's vessels. The attestation, the celarations and the interrogations put in legal form by competent authorities will serve you as title to claim the payment of the sums which may ultimately be due to you.

Since the navigation which you are about to undertake is absolutely different from others, and also the form of warfare which the *Nautilus* is intended to make upon the enemy, it is not possible to indicate in advance a fixed method of affirming the truth of the facts.

But it will be supplied by the information of the Commissary of the English Government, and by the Maritime Prefects every time it becomes necessary.

(signed) FORFAIT.

Fulton accepted the conditions even though he had already spent more than 10,000 francs of his own and others' money (mostly Barlow's). Evidently he decided he would never win a better deal, and he optimistically hoped that he would earn so much in prize money that he and his friends would recoup their expenditures many times over.

The First Consul appointed still another commission to report on the new experiments. Fortunately for Fulton, his friends Monge and LaPlace were named to it, plus another sympathizer, Constantine Francis de Chasseboeuf, Comte de Volney, a scholar, author, and traveler. With a friendly commission, Fulton began to refurbish the *Nautilus*. He moved it from Le Havre to Brest and made his first plunge on July 3, 1801. He had enlarged the boat to accommodate four men. He wrote this enthusiastic report to the commission: [14]

Paris 22^d, fructidore An 9

Robert Fulton to the citizens Monge, LaPlace and Volney, members of the National Institute, and Commissioners appointed by the first Consul to promote the invention of Submarine Navigation—

Citizens, yesterday on my return from Brest I received your note, and will with pleasure communicate to you the result of my experiments, during the summer, also the mode which I conceive the most effectual for using my invention against the enemy. Before I left Paris I informed you that my plunging boat had many imperfections, natural to the first machine of so difficult a combination, added to this I found she had been much injured by the rust during the winter in consequence of having in many places used Iron bolts and arbours instead of copper or brass, the reperation of those defects and the difficulty of finding workmen consumed near two months And although the machine remained still extremely imperfect yet she has answered to prove every necessary experiment In the most satisfactory manner.

On the 3d of thermidor I commenced my experiments by plunging to the depth of 5 then 10 then 15 and so on to 25 feet but not to a greater depth than 25 feet as I did not conceive the Machine Sufficiently Strong to bear the Pressure of a Greater column of water. At this depth I remained one hour with my three companions and two candles burning without experiencing the least inconvenience.

Previous to my leaving Paris I gave to the Cn. Gueyton member of
the Institute a calculation on the number of cube feet In my boat
which is about 212 in Such a Volume of Air he calculated there
would be sufficient Oxizine to nourish 4 Men and two small candles
3 hours. Seeing that it would be of great Improvement to despence
with the candles I have constructed a Small window in the upper
part of the Boat near the bow which window Is only one inch and a
half diameter and of Glass 9 lines thick, with this prepared I descended on the 5th of thermidor to the depth of between 24 and 25
feet at which depth I had Suffecient light to count the minutes on
the Watch, hence I conclude that 3 or 4 such windows arranged in
different parts of the boat would give suffecient light for any operation during the day each window may be Guarded by a Valve in
Such a manner that Should the glass break the Valve would immediately Shut and Stop out the Water, finding that I had air and
light Suffecient and that I could Plunge and Rise perpendicular
with facility. On the 7th Ther^d I commenced the experiments on
her movements. At 10 in the Morning I raised her anchor And
hoisted her Sails which are the Mainsail and Gib the breeze being
light I could not at the Utmost make more than about two thirds of
a league per hour. I tacked and retacked tryed her before and by
the wind And in all these operations found her to Answer the helm
And Act like a common dul Sailing boat, After exersising thus
About An hour I lowered the mast and Sails and commenced the
operation of Plunging this required about two Minuets. I then
placed two men at the engine which gives the Rectileniar Motion,
And one At the helm, while I governed the machine which keeps
her ballanced between two waters. With the bathomater before me
And with one hand I found I could keep her at any depth I
thought Proper the men then commenced movement and continued about 7 Minuets when mounting to the Serface I found we
had gained 400 Matres. I again plunged turned her round under
water and returned to near the Same place. I again plunged And
tried her movements to the right and left, in all of which the helm
answered. And the compass acted the same as if on the serface of
the Water having continued these experiments the 8, 9, 10 and
12th until I became fameliar with the movements And confidence
in their operation, I turned my thoughts to Increasing or preserv-
ing the Air, for this purpose the Cn. Gueyton advised to precipitate
the carbonic acid with lime, or to take with me bottles of Oxizine
which might be uncorked as need required; but as any considerable
quantity of bottles would take up to much room, And as oxizine
could not be created at Sea without a Chymical operation which
would be Very Inconvenient, I adopted a mode which occured to

me 18 months ago which is a Simple Globe or bombe of copper capable of containing one cube foot to [Manuscript is torn here] A Pneumatick Pump by means of which Pump 200 Atmospheres or 200 cube feet of common Air may be forced Into a Bomb consequently the Bomb or reservoir will contain As much oxegine or Vital air as 200 cube feet of common respirable Air, hence if according to Cn. Gueyton's Calculation 212 feet which is the Volume of the boat will nourish 4 Men and two small candles 3 hours this additional reservoir will give Suffecient for 6 hours—this Reservoir is constructed with a measure and two cocks So as to let measures of Air Into the Boat as Need may require—

Previous to my leaving Paris I gave orders for this machine but it did not arrive till the 18 of thermidore on the 19 I ordered 2 Men to fill it which was an operation of about one hour. I then put It into the boat and with my three companions but without candles plunged to the depth of about 5 feet, At the expiration of one hour and 40 Minuets I began to let off Measures of air from the reservoir and So on from time to time for 4 hours 20 Minuets without experiancing any Inconvenience—

Having thus succeeded
    To Sail like a common Boat
    To obtain Air and light
    To Plunge and rise Perpendicelar
    To turn to the right and left at pleasure
    To steer by the Compass under Water
    To renew the Common Volume of Air with facility

And to Augment the respirable air by a reservoir, which may be obtained at all time, I conceived every experiment of importance, to be proved in the most satisfactory manner hence I Quit the experiments on the Boat to try those of the Bomb Submarine. It is this bomb which is the Engine of destruction the Plunging boat is only for the purpose of carrying the bomb to where it may be used to Advantage. They are constructed of Copper and of different sizes to contain from 10 to 200 Pounds of powder each bomb is arranged with a Gun lock in Such a manner that if it Strikes a Vessel or the Vessel Runs against it, the explosion will take place and the bottom of the Vessel be Blown in or so Shattered as to insure her destruction. To prove this Experiment the Prefet Maritime, And Admiral Vellaret ordered a Small Sloop of About 40 feet long to be anchored in the Road, on the 23d of Thermidor With a bomb containing about 20 Pounds of powder I advanced to within about 200 Matres then taking my direction So as to pass near the Sloop I

Struck her with the bomb in my Passage the explosion took Place and the Sloop was torn into Atoms, in fact nothing was left but the buye and cable, And the concussion was so Great that a Column of Water Smoak and fibres of the Sloop was cast from 80 to 100 feet in Air, this Simple Experiment at once Proved the effect of the Bomb Submarine to the Satisfaction of all the Spectators; of this experiment you will See Admiral Villarets description in a letter to the Minister of Marine—

Having Given in a Short Sketch of the Succession of my Experiments, the mode of using these inventions Against the enemy is now to be considered, on this Point time and experience will make numerous improvements As in all other new inventions and discover modes of operation which could not possibly accur to me; when Powder was Invented Its Infinite applications were not thought of, nor did the Inventors of the Steam Engine conceive the numerous purposes to which It could be applied, in like manner it is Impossible at present to See the Various modes, or the best methods of Using a plunging boat or the bomb Submarine—

But as far as I have Reflected on this point I conceive the best operation to be as follows—

First

To construct one or two Good Plunging Boats each 36 feet long and 12 feet wide Boats of this capacity would be Sufficient to contain 8 men and Air for 8 hours. With Provisions for [paper is torn here] and transport from 25 to 30 Bombs at a time, their Cylenders Should be Bradd and of a Strength to admit of descending 60 or 80 feet under Water in case of need And they may be Constructed to Sail from 5 to 7 Miles per hour; here it may be well too proove that Quick Sailing is not one of the most important considerations in this invention, if such a boat is Pursued, She plunges under water and as She Can remain under Water from 4 to 8 hours and Make at least one Mile Per Hour She Could rise Several miles from the Place where She Plunged to renew her air, thus the enemies Ports could be approached, And particularly under the cover of the Night Nor do I at Present See that any Possible Vigilence could Prevent these invisible engines entering their Ports and Returning at Pleasure—

Second

Let there also be Some hundreds of Bombs Submarine Constructed of Which there Are two Sorts one arranged with Clockwork in Such a manner as to Go off at any Given Period from 4 Minuets to

4 hours, the Other with a gun Lock as before mentioned So as to go off when it Strikes against a Vessel or when a Vessel runs Against it. Each of these carcasses is arranged So as to float from 4 to 15 feet under water in Proportion to the Water which the Vessels to be attacked Draws, And in this there are two advantages, the first is that the bomb is Invisible, the Second is that when the explosion takes place under water the Pressure of the colume of water to be removed forces the whole action of the powder Against the Vessel; it was the resistance of the water which caused the Sloop on which I proved the experiment to be reduced to Atoms; for Water when Struck Quick such as the Stroke of a cannon ball or the expansion of Powder acts like a Solid, and hence the whole force was Spent on the Sloop or rather passed through the Sloop in finding its Passage to the air by the perpendicular and Shortest line of Resistance—the Same effect would no doubt be produced on a Vessel of Any dimensions by applying a Proportionate Quantity of powder Such as 2, 3 or 4 hundred Weight.

Therefore being prepared with plunging boats and Bombs submarine let the business of the boats be to go with cargoes of bombs and let them loos withe the current into the harbours of Portsmouth, Plymouth, Torbey or elsewhere, those with their graplings floating under water could not be perceived Some would hook in the cables, bow or Stern, or touch in their Passage; many no doubt would miss but Some would hit go off and destroy the Vessels they touched, one or more Vessels Destroyed in a Port by such invisible agents would render it to dangerous to Admit of any Vessel remaining. And thus the enemy may At all times be attacked in their own Ports—and by a means at once cheap, Simple And I conceive certain in its operation. Another mode Should be to go with cargoes of Bombs and Anchor them in the entrance of rivers So as to cut off or Blockade the commerce 2 or 3 hundred for example Anchored in the Thames or the channels leading to the Thames would completely destroy the commerce of that river and Reduce London and the Cabinet of St. Jameses to any tirms; no Pilot could Steer clear of Such hidden dangers, no one dare to raise them even if hooked by graplings as they could not tell the moment they Might touch the Secret Spring which would cause the explosion and destruction of everything Around them. No Vessel could Pass without the utmost danger of running on one of them. And her instant destruction, if this measure Should ever become necessary Some Vessels Will most certainly be destroyed and their Destruction alarm the whole commerce of the Thames, by this means the Thames may be blockaded and the trade of London completely stoped nor can the combined fleets of England prevent this Kind of attack—And this is Perhaps the most Simple and certain means of convincing England that Science can put her in the Power of France and of compelling her to become a humble Pleader for the liberty of the Seas She now denies to her Neighbors—I therefore conceive that it will be good Policy to commence as Soon as Possible the construction of the Boats and bombs if they can be finished before the arrival of Peace their effects may be Proved during this War Should Peace be concluded before they are finished the experiments can be continued Men can be exersised in the use of the engines; And it is Probable in a few years England will See it her best policy never to give France reason to exersise this invention against her—if England cannot prevent the Blockade of the Thames by the means of plunging boats and Bombs submarine, of what use will be her boasted navy, the free Navigation of the Thames nourishes the immense commerce of London And the commerce of London is the Nerves and Vitals of the Cabinet of St. Jameses—convince England that you have the means of Stopping that Source of Riches—And She must Submit to your terms—

Thus Citizens I have presented you with a Short account of my experiments and Plan for using this invention Against the enemy, hoping that under your protection it will be carried to Perfection, and Practised to promote the Liberty of the Seas—

Health and Sincere Respect
ROBERT FULTON

The inventor later uses the word "torpedo" for submarine bomb. Fulton borrowed the term from the fish of that name (also called the electric ray). Actually, we would call his "torpedo" a floating mine today.

After the successful trials with the modified *Nautilus,* Fulton did something that probably ruined his chances to go into full partnership with Napoleon. He dismantled his boat, claiming, "I am sorry that I had not earlyer information of the Consuls desire to See the Plunging boat, when I finished my experiments. She leaked Very much and being but an imperfect engine I did not think her further useful." [15]

Why the haste? Fulton feared that the French would steal his plans and ideas and never pay him a franc. Earlier, he had refused to exhibit his drawings for the same reason. This reluctance had accounted, in part, for the years of delays, but he had stubbornly persisted.

Another reason why he may have scrapped the *Nautilus* lies in the fact that he recognized she would never prove practical without some means of mechanical power. In 1801 he began experimenting with surface vessels by which he could deliver his torpedoes. He fitted a pinnace with a screw propeller driven by hand cranks. Even with fifteen sailors straining at the cranks, he never achieved more than four miles per hour. He asked the local French admiral for protection with six armed cutters to guard his escape. The admiral refused, wanting nothing to do with such a barbarous method of warfare. Fulton bravely tried to deliver the torpedoes without protection. Again, the British brigs disappeared, probably warned of the new developments by the Secret Service—lookouts had watched the sea with glasses and small boats patrolled around British ships near shore.

Fulton followed Bushnell's thinking and experiments every step of the way. The earlier inventor had also concluded that the submarines could not deliver the torpedo to the enemy hull and had experimented with surface means, including the floating of the mines toward the British ships. Fulton developed refinements. He thought of the idea of anchoring the mines in the expected path of the British vessels and of using the submarine to lay them. Even these improvements didn't impress Napoleon.

Fulton returned to Paris from Brest, again expecting to hear almost immediately from Bonaparte. Nothing happened. He wrote a personal letter to him, urging and begging favorable action. Again, no reply.

When the American heard second-hand that the First Consul had spoken of him as a charlatan and a swindler, intent only on extorting money,[16] he began to realize that he had no hope of success with the submarine in France.

There is evidence that the submarine did raise a flicker of interest in Napoleon as late as 1803, but the flicker died and Fulton never knew about it. On July 21 of that year, Bonaparte had written to an official in the Marine Department: "I have just read the proposition of Citizen Fulton that you have sent to me much too late *to permit it to change the face of the world.* However, I desire that you will immediately refer its examination to a commission composed of members chosen from the different classes of the Institute. It is there that the wisdom of Europe should seek judges to solve the problem in question, As soon as the report is made it will be transmitted to you and you will send it to me. Be sure that this will not take more than a week." [17]

Evidently Napoleon had forgotten earlier investigating commissions and his views on Fulton. The latest commission may have vetoed the submarine because the French archives contain nothing more on Fulton or the *Nautilus*.

But the recognition of failure in France dawned slowly on the inventor. While waiting in Paris, he dabbled in still another project. In 1801, Robert R. Livingston, chancellor of New York State, statesman and one of the wealthiest men in America, arrived in France as American Minister to the French Government. Barlow welcomed him, if only to interest Livingston as an investor in the submarine because the project had developed into an excessive financial drain. Livingston was not interested in vessels that went under the sea, but he was interested in one that could ply the surface of the world's waters—the steamboat. We shall postpone for a while the account of Fulton's adventures with this invention in Paris.

Instead, let's pursue the story of the submarine, this time to England.

8 | THE ULTIMATE WEAPON

From about 1800 on, Fulton had been surreptitiously supplying friends in Britain, principally Lord Stanhope, with tidbits about the submarine. Fulton himself and some of his early biographers defended such near-treason on idealistic grounds. Because his method of submarine warfare would inevitably destroy all navies, what difference would it make which country introduced it if he achieved the ultimate goal of bringing peace and free trade to all the world?

Fulton evidently felt the weakness of his position because he also offered other arguments: Napoleon's despotism; the failure of France to pay him as had allegedly been promised; the end of Republicanism in France; the realization that the English government, which he had formerly described as "monstrous," had somehow turned into a democracy at heart.

Of course, he did not reveal publicly another reason—money. Fulton in his own private writings, makes it clear that he never seriously thought the British would actually use his submarine-torpedo system because it would be to their disadvantage to do so. Any cheap way to cripple overwhelming seapower would harm England; therefore the canny British merely wished to control the invention. Fulton knew this, as this extract proves from a June 30, 1804 letter to "Mr. Smith," [1] a British agent charged with luring Fulton to England:

> It will be seen that England may draw many advantages from these inventions, or that they may be turned to the total destruction of the British marine. In either case, it is of importance to the British

government to have the entire command of such engines to do with them as they think proper. But as these inventions are the produce of my labors for some years, I now consider them as rich gems drawn from the mines of science and which I and my friends have a right to convert to our own advantage, and which I now offer for sale to the British government.

For his designs and explanations of attack methods, "I require the sum of £100,000." [2] That would be about $3.6 million in today's purchasing power.

Did Fulton seriously think Britain would pay such sums for a flimsy craft that had never sunk so much as a rowboat? He had exaggerated the prowess of the *Nautilus* in his letters to Stanhope. Furthermore, British agents had themselves wittingly or unwittingly made his weapon sound much more formidable than it really was. For example, one Evan Nepean on June 19, 1803, sent a secret circular to commissioners of the Admiralty, claiming that Fulton had been underwater seven hours, traveled at underwater speeds of two and a half miles per hour and destroyed a vessel off Brest with only twenty-five pounds of gunpowder.

Even if it was largely a phantom, Fulton had a bargaining tool. Yet the British had one too. The inventor wanted to buy a Boulton & Watt steam engine for the proposed steamboat. The British dangled the possibility that he might be permitted to do so—if he came to England.

This bait lured Fulton to Amsterdam in the fall of 1803 where he was to meet "Mr. Smith," the agent sent at the instigation of Stanhope. Fulton describes the incident: "I agreed on certain conditions and Mr. Smith set off for London to give in my terms. I then met him in Amsterdam in December with the reply which, not being satisfactory, he returned to London with other proposals and I went on to Paris." [3] Beginning in December, Fulton waited in Amsterdam for three months, but "Smith" never came. The inventor revealed the state of his nerves by a series of atypical caricatures of the Dutch which he did during that period to while away the time. We are told they were savagely sardonic, totally different from anything else he ever did.

He returned to Paris with no news, all his projects in limbo. When he had almost lost all hope, "Smith" suddenly appeared at 50 Rue de Vaugirard. Imagine the scene: Looking nondescript and insignificant in the manner of secret agents since the world began, Smith insisted that the two go behind closed and locked doors, draw the drapes or blinds, and only then revealed his documents that had been concealed on his person. The letter, in cipher, bore the signature of Lord Hawksbury, the Home Secretary. Decoded, it informed the inventor:

> The responsibility attached to His Majesty's ministers in their official capacity renders it impossible for them to advance the sums which you have required, in the form pointed out by you, without exciting such public attention as must be equally unpleasant to you and His Majesty's ministers; if, however, you have sufficient confidence in His Majesty's government to offer them your invention, you may rely on being treated with the utmost liberality and generosity . . . A negotiation personally conducted would smooth many difficulties, and every facility and protection shall be granted to you. And should you be disposed to accept active employment from the British government, you may rely on the most liberal treatment, proportioned to your efficient service.[4]

Although this offered nothing specific, Fulton accepted it because he had reached the end of the line. He needed the Boulton & Watt engine for the steamboat. He had lost all hope that the French would do anything about the submarine-torpedo plan. One of his last acts before quitting France was to send his manuscripts and drawings to the United States. Unfortuntely, the vessel carrying them was wrecked. Although his case was recovered, the contents were so waterlogged that most of it was ruined. This accident, together with Fulton's secretiveness, accounts for the gaps in his story up until 1804.

He left Paris on April 29, 1804 and reached London by a circuitous route on May 19, traveling as "Mr. Francis." Fulton would use that name most of the time during his stay in England. Although a thin disguise, it amused him because intrigue was part of his nature, and because the name was the anglicized form of the name that the Duchesse de Gontaut-Biron had used when he had met her crossing from England to France.

When he went around to Whitehall, Fulton at once ran into a snag. The government with which he had negotiated had left office. William Pitt was now Prime Minister. Lord Melville was First Lord of the Admiralty. They put him in touch with someone he named "Mrs. Hammond" (perhaps Lord Hawksbury since Fulton had a fondness for using pseudonyms which had the same initial as the real name of the subject).

Fulton thought he would settle his English affairs in a few weeks. How little he knew the English capacity for delay! Now that they had him in England, they intended to keep him there, never denying flatly any of his demands, never wounding his amour propre, even putting up with his near-insults which normally would have ruined any chance of a business deal in those aristocratic days. For example, on June 22, after he had been in England a little more than a month, he wrote "Hammond":

> The first day I had the pleasure of seeing you, I promised you candor, and should time make me more known to your government, they will find frankness one of the leading lines of my character. Now I candidly declare that having been here five weeks in some degree like a prisoner, and at present as much in the dark as on the day of arrival, such a state of suspense begins to grow extremely unpleasant.[5]

He had been enticed to England by "flattering and I believe candid promises . . . and as yet I do not repent it . . . I came here to acquire wealth by communicating a new system to the government." He required a reply by the next Tuesday—an unheard of demand to the government from a mere commoner who was not even an Englishman.

From the references to "prisoner," we can deduce that Fulton had been kept at least incommunicado, if not actually locked up, since his arrival. The British were taking no chances with their catch.

In terms of bureaucratic speed, five weeks was actually surprisingly fast. Fulton's written specifications had gone to a distinguished commission: Sir Joseph Banks, president of the Royal Society; Henry Cavendish, the great chemist; Major William Congreve, artillery officer and rocket inventor; Sir Home Popham, naval commander; and John Rennie, probably the greatest civil engineer in England. They examined a plan similar to the one Fulton had refused to show Napoleon without payment two years earlier.

A copper cylinder, six feet in diameter and eighteen or twenty-four feet long, fitted with a dome for entering and exiting was put in an ordinary sailboat thirty feet by ten. The extra space in the enveloping ship contained the water tanks for submerging, as well as racks holding thirty bombs, which were to be anchored in channels or permitted to drift into enemy harbors. In short, this was an enlarged *Nautilus*, redesigned to carry more bombs, to sail more efficiently, and to range farther.

The commission decided that the submarine was mechanically possible, but unlikely to achieve much practical result. Yet they did recommend an alternate plan Fulton had suggested—to blow up fleets in enemy harbors with submarine bombs manipulated from partially submerged rafts carried close to shore defenses by ordinary vessels.

The inventor probably at first regretted that he had thrown in the alternative because he, of course, wanted his pet, the submarine, to have won recommendation. Yet the commission saw, as most other knowledgeable people also had seen, that hand-crank propulsion would never work effectively in combat. He dashed off a hot-tempered note that reminded the government that it could not suppress his invention without paying him. He also warned that if their decision "should necessitate me to seek fortune elsewhere," so much the worse for Britain.

The Pitt government countered by suggesting that it might use the alternative. That didn't mollify Fulton because such a decision meant he might have to wait months to be paid, or even years until the surface delivery system was proven. He had hoisted himself on his own petard by implying in his prospectus that he had already subjected the surface method to extensive tests. Actually, he only had experimented with it cursorily a few times at Brest in 1801.

That the British thought Fulton important—or perhaps dangerous—is also indicated by the fact that Prime Minister Pitt himself invited Fulton to breakfast on July 20. The American's report of the meeting says that when Pitt "remarked that this is a extraordinary invention which seemed to go to the destruction of all fleets, I replied that it was invented with that in view, and as I had no design to deceive him or the government, I did not hesitate to give it as my opinion that this invention would lead to the total annihilation of the existing system of maritime war." [6]

Perhaps Pitt smiled because he already knew what it has taken another 170 years and two world wars to teach: There is no such thing as a war to end wars or the ultimate weapon. He pointed out gently: "But in its present state of perfectionment those who command the seas will be benefited by it, while the minor maritime powers can draw no advantage from what is now known." Fulton countered by insisting on the deadliness of the submarine. Pitt, the nimble-brained parliamentary debater, rebutted with these arguments:

> That it would probably be some years before any nation could bring into perfection such a vessel; that it was not to the interest of the British government to use such vessels; that consequently there was not at present much danger to be apprehended from that part of my system; at all events there would be time to fit future politics to future circumstances . . .[7]

Pitt offered to retain Fulton to perfect a surface delivery system for mines, including a salary of £200 a month (about $7,200 in today's purchasing power) and £7000 ($252,000) for expenses. If the government decided to suppress his invention, he would receive £40,000 ($1.4 million) provided an arbitration committee agreed that his weapon was practical. If the government used the invention, he would superintend operations and receive half the value of the ships destroyed. If he withdrew from active participation, his share would drop to one-fourth. Fulton could not divulge any part of his principles and methods for fourteen years and had to promise to withdraw an account of his method which he had confided to a friend (probably Barlow) in case he died. Fulton signed the agreement, but he nevertheless did not withdraw the account entrusted to the friend.

The sum of £200 as a monthly salary was shrewdly conceived. It failed to approach his extravagant demands, but it also was a princely amount to the chronically impoverished Fulton. It would enable him to live well, as he ardently wished to do. It was the largest regular income he had ever received in his life. No wonder he accepted it!

If his system sank any French ships, he could become a millionaire in terms of modern purchasing power. To snare such a fortune, the inventor turned from his motto: "The freedom of the seas will be the happiness of the world." Instead, he would help increase the blockading power of the British fleet by attacking the French flotilla bottled up in Boulogne harbor.

Was Fulton an unprincipled opportunist? Perhaps. But he was also a thirty-nine year old man, almost desperate to succeed dramatically in something. Thus far, he could point only to failures and half-successes. At this time, he suffered from stomach trouble, insomnia, and nightmares when he did fall into a fitful sleep. These stemmed from his years of frustration. He had a self-esteem of almost megalomaniacal proportions, yet none of his projects had fully succeeded. Once he probably sincerely believed in his abstract motto. Now he had to have some tangible evidence—money—of his own worth.

And the British at this time were also clutching at straws—because of the Napoleonic threat of invasion, dubbed The Great Terror by English historians. Napoleon chose Boulogne as the base for the French operations; along the coast on either side of the harbor, an army of some 150,000 men was encamped, and specially constructed flat-bottomed boats for their transport were brought from all parts of France. The army received constant practice in embarkation and disembarkation. While they knew much about the French plans, the blockading British could do little damage to this armada because the shallow water on the coastline prevented the entry of their big warships.

Fulton's scheme was one among many that the English government considered on how to reach into Boulogne harbor to destroy or at least damage the French. In the meantime, the English male citizenry along the southern coast organized into local militia to protect their shores.

To use his barges and army, Napoleon had to win at least temporary control of the English channel. French fleets were stationed at Brest, Rochefort, L'Orient, and Toulon; all of them were watched by British blockading squadrons which had to guard against the French sneaking past them.

So, the British felt great pressure, which goes far to explain the generous terms granted to Fulton. He set to work at once to prepare for an assault on the Boulogne flotilla, an attack known as the "Catamaran expedition." The submarine "carcasses," "bombs," or "coffers" used by Fulton were similar to those he had experimented with in France. An officer who participated in the operations described them as square with wedge-shaped ends "of thick plank lined with lead. A plank is left out for filling it. When filled the plank is put in, nailed and caulked, paid all over with tar, covered with canvas, and paid with hot pitch."

Clockwork, set to run a certain time before releasing a hammer, was attached to the carcass. The clockwork was started by the removal of a pin. The coffers, some of which weighed two tons and were eighteen feet long, were loaded with shot so as to float just awash and thus escape observation. To each coffer were attached two lines, floated with pieces of cork, one a tow line and the other a grapnel. The latter was intended to be hooked on the cable by which a barge was riding at anchor, when the coffer would swing round by the tide and lay alongside.

The coffer's delivery system was a catamaran consisting of "two pieces of timber about nine feet long and nine inches square placed parallel to one another," but far enough apart to allow men to sit between them on a bar. The sailors were nearly flush with the water, to minimize the chance of being seen. Dressed in black, including black masks, they were to be dropped off by a normal boat close to shore, then paddle to shore, attach the mine, set the clockwork going, then paddle back to the mother ship as fast as possible. Many such catamarans with their deadly tows were launched on the night of October 2, 1804.

William Cobbett described the expedition in his anti-administration *Political Register*.[8] The Dundas referred to is William Dundas, Lord Melville, first lord of the Admiralty. The Earl of Harrowby is Dudly Ryder, Pitt's Secretary of State for foreign affairs.

Dundas is gone to Boulogne;
He has a *pawky* plan
To burn the French flotilla:
'Tis called *Catamaran* . . .

Like ladies in romances
Their knights' exploits to spy
Aloft in Walmer Castle
Stand Pitt and Harrowby . . .

Dundas our tars haranguing
Now shows his new-made wares.
As at some peddler prating,
Jack (Tar) turns his quid and swears . . .

"See here my casks and coffers
With triggers pulled by cocks."
But to the Frenchman's riggings
Who first will lash these blocks?

"Catamarans are ready"
(Jack turns his quid and grins)
Where snugly you may paddle
In water to your chins.

"Then who my blocks will fasten
My casks and coffers lay?
My pendulums set ticking
And bring the pins away?"

"Your project new?" Jack utters;
"Avast. 'Tis very stale;
'Tis catching birds, landlubbers!
By salt upon the tail!"

So fireships, casks, and coffers
Are left to wind and tide;
Some this, some that way wander,
Now stern before, now side.

Ships, casks, and coffers blazing
Now brings Vauxhall to mind;
As if ten thousand galas
Were in one gala joined.

Aloft in Walmer Castle
Stand Pitt and Harrowby,
"The fireworks are beginning!"
With eager joy they cry.

"There in that blaze go 50!
And there go 50 more!
A hundred in disorder
There run upon the shore!"

From them the joyful tidings
Soon flew to London town:
By hundred and by thousands
They burn, sink, kill and drown.

How long Dundas for morning
His triumphs to survey,—
But lo. the French are laying
Just where before they lay . . .

But now them who never
Did England's hopes deceive,
Our soldiers and our sailors,
Their business let us leave.

May Pitt from colonelling
Retire upon half pay
And Admiral Lord Melville
The yellow flag display!

Cobbett derided the expedition because it failed dismally. It sank nothing, the bombs exploding harmlessly on the shore. Did such an outcome send Fulton into fits of despair? Perhaps privately, but publicly he rose to new heights, as he frequently did to face reverses. He mustered all his persuasive skills, convincing Pitt that the expedition had foundered because the officers commanding it had not followed his instructions. The Prime Minister put up another £3,000 to build more bombs. A second attempt was made on the night of December 8, 1804, this time commanded by a Fulton sympathizer, Captain Sir Home Popham. He sent an explosion vessel and two carcasses against Fort Rouge in Calais Harbor. The vessel exploded but one carcass did not go off and the other had to be brought back.

Although this episode went better than the Boulogne affair, it was no unqualified success. Nevertheless, the British awarded Fulton an additional £10,000, "to relieve me of some pecuniary embarrassments." The payment discloses two facts: That the British still wanted to keep a string on the inventor and that when he had money he spent it.

While cooling his heels waiting for the British government to act, the inventor became a patron of the arts and a man about town. One evening at the opera, he spied the French charmer whom he thought of as Mme. François or Mme. Montaut. He stared so unselfconsciously that she noticed him and acknowledged his presence with a deep bow. He rushed to her box, oblivious of the fact that she sat with the Duke of Portland and Lord Clarendon. Grasping her hand, he exclaimed, "What a pleasure, Mme. Montaut, to find you again here! I could hardly believe my eyes."

A Frenchman corrected him. "Monsieur is in error, for Madame is the Viscountess of Gontaut." All but Fulton laughed.

The Viscountess recalls that Fulton next said, "Oh, dear me!, This is too much! Always changing her name. It is enough to drive one mad. But I see that these gentlemen are in on the mystery. If it is a joke, let us laugh together."

The lady reported in her memoirs: [9] "His good and simple manner touched me, and I told him that since we were in a friendly country, I could explain the mystery." At last, Fulton learned the true story.

"Now I understand," said the inventor. "I wish to compliment your husband on having a wife who was on the point of sending me to the devil."

Another romantic episode involved a young English widow. All we know about her is that she was rich, named Clarissa, and enjoyed ice skating. That's how the pair met. Their friendship bloomed all through 1805, to the extent that early in 1806, Robert wrote Barlow that he intended to marry her. Barlow replied:

> I write you with a heavy heart. Your letter of the 12th of January came upon us like a shipwreck. We see in it at least the wreck of our most brilliant projects of domestic happiness, if not of public usefulness . . . We can say nothing to your proposal except that you ought by all means to pursue your own ideas of your own happiness, well weighed and well considered. On this last clause I must offer a word, tho' it may probably come too late to be of use. My friendship is unlimited and unabated, and I have no reason to doubt of the variety of excellence you find in the person you describe. But her education, habits, feelings, character and cast of mind are English and London. And what is perhaps most unfortunate for you, she has a fortune. These things render it extremely improbable that she can be happy in this country. I should think it equally impossible that you can be very happy in that country. Your mind is American, your services are wanted here. Your patriotism, your philanthropy, your ideas of public improvement, your wishes to be a comfort to me and my wife in our declining years (if we should unluckily have many of them) would tend to make you uneasy at such a distance from the theatre of much good.[10]

The letter may have given him second thoughts. In any event, Fulton didn't marry her, and we never hear another word about Clarissa.

Fulton also spent some of his time and money on paintings, buying Benjamin West's *Ophelia* and *King Lear* for 125 and 205 guineas, respectively. He intended that these paintings form the nucleus of a Gallery of Fine Arts to be set up in Philadelphia.[11] What better painter to begin the collection than the native Pennsylvanian, West? This is the first case on record of the practice, now so common, of the purchase in Europe of original works of art for America.

Although a generous gesture and characteristic of one facet of his nature, Fulton did not want his actions to go unnoticed. He wrote to publications in England and America about his purchases. His skills as a publicist were rewarded; he won attention on both sides of the Atlantic. (*King Lear* eventually found its way to a Boston museum. *Ophelia* is lost.)

Another artistic project involved his friend Barlow, who rewrote his youthful *Vision of Columbus* into the much more elaborate *Columbiad,* a poetic history of America. Fulton, a better engineer than literary critic, encouraged his friend to bring out a sumptuous edition befitting his affluence and status as poet of America's rising power. The inventor, who seldom did anything by halves, planned a large illustrated volume. A trial "sketch" for the book was done by John Vanderlyn, an American who had previously done their portraits. They had asked him for a scene from the American wilderness. His version, now known as *The Death of Jane McCrea,* is a full-blown painting of a civilized dell containing not Indians, but half-naked giants drawn from classic sculpture. Their victim is a damsel in French afternoon dress with such extreme decolletage that her breasts are nearly bare. Fulton and Barlow rejected the sketch, and Fulton drew the sketches for the book himself.

Barlow spent $1,000 to commission paintings from Fulton's sketches by a prominent artist, Robert Smirke. Fulton added $5,000 of his own to have the paintings engraved. The eleven illustrations, plus a frontispiece of Barlow's portrait engraved from Robert's own painting, went into the book. Fulton had the plates made and the prints struck off in 1805. He brought them with him to the United States.

Although the volume didn't appear until 1807, it was the most beautiful book yet manufactured in America. The poem has been forgotten, yet the book deserves an honored place in the history of American graphic arts. In addition to the illustrations, it was printed in a handsome, eighteen-point, specially made up type on fine paper in quarto size and bound in leather with gold stamping. Even the caustic Francis Jeffrey in the *Edinburgh Review* said he never had seen a more attractive book published in England. He conceded: "The infant republic has already attained to the very summit of perfection in the mechanical part of bookmaking." [12]

Barlow acknowledged Fulton's role in the edition's manufacture and dedicated it to him: "This poem is your property. I present it to you in manuscript, that you may bring it before the public in the manner you think proper . . . You designated the subjects to be painted for engravings; and, unable to convince me that the work could merit such expensive and splendid decorations, you ordered them to be executed in my absence and at your own expense." [13]

Perhaps the ostentatiousness embarrassed Barlow. How could a Jeffersonian Republican who embraced the simple things of life be associated with an extravagant book that cost $10,000 ($70,000 in modern purchasing power) to publish and was to be sold for $20 ($140) a copy? He solved his dilemma by placing the blame on Fulton who, although also professing to be a Jeffersonian, really was no more modest than a Napoleon. (Fulton affected modesty only when he thought it would impress someone.)

Although Fulton dabbled in artistic and social pursuits for much of 1805, he never forgot his main purpose: To become rich. He bombarded various government officials with letters and visits urging either that they pursue his torpedo scheme or pay him to drop it. He didn't care which.

A letter of August 9, 1805 to Pitt contains a typical blend of commercialism, idealism, and veiled threat:

> I proposed to take the value of one ship of the line of £100,000 to let the discovery lie dormant . . . Being a neutral . . . it was not my intention to take part in European wars . . . If it is an invention which is capable of working a total revolution in marine war *and which I believe,* I of course must have a high idea of its value to myself and country. [14]

Even if the British ministers did not take seriously his ambiguous reference to moving his project to America, they decided to try out the torpedo again—undoubtedly because The Great Terror continued. By late summer of 1805, intelligence reports indicated that Napoleon had collected 1800 invasion barges in Boulogne harbor and had increased his army to 200,000 soldiers encamped there. Bonaparte had taken personal command and now stayed in the town.

Nelson himself had attempted a sea assault on the harbor, but the shore defenses were too much for him. A desperate Pitt returned to Fulton. He collected another £4400 to build more torpedoes. On the night of September 30, 1805, speedy galleys, each with a commanding officer, eight oarsmen, and a coxswain, slipped into Boulogne harbor and in two instances attached a yoke of two torpedoes to the anchor cables of French gun brigs. Rowing away under musket fire, the crews heard tremendous explosions in the harbor. Torrents of water shot up, but alas!, when the seas calmed, the French vessels remained serenely at anchor. In reporting on this affair, the French mentioned that the next morning they found on shore "a lock like that of the fire machines which the English used last year with so much ridicule and with so little success."

Fulton reported laconically on this latest failure: "I was much at a loss to account for the brig not being blown up," but a half hour's thought gave him the answer—the torpedoes had been so much heavier than the water that they were not washed under the keel but had hung perpendicularly from the sides of the boat.

Now an expert at the psychology of persuasion if not in blowing up French boats, Fulton knew he had to restore confidence in his invention. He won permission to try to sink a captured Danish brig lying in Deal Harbor. Skeptical English Navy men arrived to watch the experiment, among them a Captain Kingston who joked that if Fulton placed the torpedoes under his cabin while he had dinner, he would continue his meal unperturbed. Moments later the brig went down in a dramatic explosion. In describing this success on October 16, 1805 Fulton is not laconic:

> Two boats, each with eight men, commanded by lieutenant Robinson, were put under my direction. I prepared two empty Torpedoes in such a manner, that each was only from two to three pounds specifically heavier than salt water; and I so suspended them, that they hung fifteen feet under water. They were then tied one to each end of a small rope eighty feet long; thus arranged, and the brig drawing twelve feet of water, the 14th day of October was spent in practice. Each boat having a Torpedo in the stern, they started from the shore about a mile above the brig, and rowed down towards her; the uniting line of the Torpedoes being stretched to its full extent, the two boats were distant from each other seventy feet; thus they approached in such a manner, that

one boat kept the larboard, the other the starboard side of the brig in view. So soon as the connecting line of the Torpedoes passed the buoy of the brig, they were thrown into the water and carried on by the tide, until the connecting line touched the brig's cable; the tide then drove them under her bottom. The experiment being repeated several times taught the men how to act, and proved to my satisfaction that, when properly placed on the tide, the Torpedoes would invariably go under the bottom of the vessel. I then filled one of the Torpedoes with one hundred and eighty pounds of powder and set its clockwork to eighteen minutes. Everything being ready, the experiment was announced for the next day, the 15th, at five o'clock in the afternoon. Urgent business had called Mr. Pitt and Lord Melville to London. Admiral Holloway, Sir Sidney Smith, Captain Owen, Captain Kingston, Colonel Congreve, and the major part of the officers of the fleet under the command of Lord Keith were present; at forty minutes past four the boats rowed towards the brig, and the Torpedoes were thrown into the water; the tide carried them, as before described, under the bottom of the brig, where, at the expiration of eighteen minutes, the explosion seemed to raise her bodily about six feet; she separated in the middle, and the two ends went down; in twenty seconds nothing was to be seen of her except floating fragments; the pumps and foremast were blown out of her; the foretopsail-yard was thrown up to the cross-trees; the fore chain plates, with their bolts, were torn from her sides; the mizen-chain-plates and shrouds being stronger than those of the foremast, or the shock being more forward than aft, the mizenmast was broke off in two places; these discoveries were made by means of the pieces which were found afloat.[15]

Influenced by this success, the British on October 27 made another attempt on the flotilla in Boulogne Harbor. Lieutenant Charles F. Payne commanded the expedition which succeeded in placing a carcass across the cable of one of the enemy's ships. The torpedo "exploded and a similar crash as the Brig lately blown up," without, however, the same destructive effect.

Soon after this, the news reached Britain of Nelson's great victory off Trafalgar on October 21, 1805. The action, while it did not end the war, gave to England the complete command of the sea. Napoleon had to relinquish all hope of invading the British Isles.

The English interest in Fulton's torpedo sank as abruptly as his plunging boat, but the inventor persisted. He bombarded various

ministries with long letters urging torpedo action to finish off the
French. After these entreaties had fallen on deaf ears for several
months, in December he changed his tune, now suggesting a final
settlement. On January 6, 1806, he had the effrontery to write
this to Pitt:

> I did not come here so much with a view to do you any material
> good as to show that I have the power, and might in the exercise of
> my plan to acquire fortune, do you an infinite injury, which Minis-
> ters, if they think proper, may prevent by an arrangement with
> me.[16]

When you consider Fulton's scant success with the submarine
and torpedoes, this letter is sheer bluff. Perhaps fortunately, for
the inventor, Pitt died in the same month and Lord Melville of the
Admiralty was soon impeached for his irregular conduct in not
keeping his private accounts separate from his public ones. So
Fulton had to deal with a new set of ministers, Lords Grenville
and Howick, who never had liked his method of warfare, and
they now saw no need for it.

Fulton's tortuous negotiations with English bureaucracy lasted
until September, 1806. He may have worried how his public
image might look if his correspondence ever became public. "I am
not a man much governed by money," he claimed. "An honorable
fame is to me a much more noble feeling." He only wanted
£40,000 ($1.4 million in today's purchasing power) "to gratify two
friends who have been kind to me and more governed by the
hope of gain than I am . . . Gentlemen, should your award not
meet their views of wealth, I shall feel free to act as I think
proper." [17]

He never revealed the identity of the two partners, and no Fulton biographers have ever learned who they were, almost certainly because they never existed.

Perhaps also with an eye to how he might appear to his own
countrymen if his letters ever became public, Fulton also threw a
dash of patriotism into his arguments:

> At all events, whatever may be your award, I shall never consent to
> let these inventions lie dormant should my country at any time have
> need of them. Were you to grant me an annuity of £20,000 a year,
> I would sacrifice all to the safety and independence of my country.
> But I hope England and America will understand their mutual in-

terest too well to war with each other, and I have no desire to introduce my engines in practice for the benefit of any other nation.[18]

As we shall see, the inventor did indeed try to interest the United States in his plunging boat and torpedoes during the War of 1812.

Finally, the British arbiters met to consider his case. They ruled that, because the invention was impractical, England did not have to pay to suppress it. So, nothing for that. They awarded him £15,000, but he had already received all but £1608.3.2 of that sum. (Even so, that remainder was more than $50,000 in today's purchasing power.)

Fulton, of course, raged, at least publicly. On September 23, 1806 he published some of his correspondence on the negotiations in a privately circulated pamphlet. The most interesting passages in it indicate that the arbiters knew about Bushnell's earlier invention. How did he react to charges that he had passed off another's ideas as his own? By attacking them, naturally—but he did it skillfully:

> That which approaches the nearest to my inventions and combinations is a machine made by Mr. Bushnel in America, during the war, which was called the American Tortoise; in which I believe he once or twice went under water; there were also some kegs of powder floated down a river with common gunlocks on them, and a string tied to the trigger, in hope that some of the British ships would take the kegs on board, and while working about them, the men might pull the string and blow themselves up.
>
> In what Mr. Bushnal did there was much ingenuity, and no one respects his talent more than I do; had he prosecuted his studies, he probably might have perfected thought after thought to the annihilation of all ships of war. But whether his mind only viewed the subject as limited to little operations, or whether he thought too many difficulties attended it, he certainly did not compose his machines so as to make them of any use, nor did he organize anything like a system; and perhaps it is for these reasons that he had abandoned the subject for more than twenty-five years, and it is now dormant in him.
>
> But imperfect and scattered ideas are so very unlike perfect engines and efficient arrangement, that the latter may with justice be called novel invention.[19]

Note that the younger man's "system" was scarcely more successful than the older man's. While Fulton did make improvements on Bushnell's basic ideas, Bushnell, not Fulton, had first developed the principles.

Fulton's first biographer, Cadawallader Colden, who knew him personally, wrote:

> Mr. Fulton did not pretend to have been the first to have discovered that gunpowder might be exploded with effect under water; nor did he pretend to be the first who attempted to apply it as the means of hostility. He knew well what had been done by Bushnell in our revolutionary war. He frequently spoke of the genius of this American with great respect and expressed a conviction that his attempts against the enemy would have been more successful if he had had the advantages which he himself derived from the improvements of nearly forty years in mechanics and philosophy.[20]

Fulton probably did make such statements when he had returned to America where many were still alive who knew about Bushnell and his experiments. Jefferson, for example, in 1813 criticized Matthew Carr's naval history of the United States for not mentioning Bushnell. However, Fulton denigrated Bushnell in Europe, especially in England.

In his pamphlet against the arbitration award, the inventor also challenged the arbiters to stay aboard an armed vessel while his "system" attacked it. They, of course, ignored him.

Privately, Fulton doesn't sound distressed about the arbitration's outcome in a letter to Barlow. After detailing his financial situation, including paying his debts and expenses, he estimated he would have £200 clear (about $7,200). He also would be allowed to ship a Boulton & Watt steam engine to America, as well as pictures worth £2,000 ($72,000). "Therefore, I am not in a state to be pitied."

Note also that Fulton says nothing about the two partners who allegedly forced him to drive a hard bargain with the British. To his best friend, who, some have speculated, was one of the silent partners, he surely would have referred to the mysterious pair—if they actually had existed. Speculation also has it that one of the partners could have been one Daniel Parker about whom we'll learn more later. However, at this time, he owed Fulton money (a novel turnabout), not the other way around.

Just before Fulton at last sailed for America in late 1806, after an absence of two decades, he wrote to Lord Stanhope and commented on his invention of stereotyping.[21] "If stereotype printing can ever succeed in this country," he said, "it must be placed on the liberal footing of other manufactures. If thrown open on a broad basis to all who choose to employ it, free from restrictions and the spirit of monopoly, it will succeed, and those who have embarked their property in it will reimburse themselves with profit."

He wrote this as he planned to set up a steamboat monopoly in New York State.

THE INVENTION THAT WOULDN'T STAY INVENTED

A practical steamboat should have been invented at least by 1790. It was the world's bad luck and Robert Fulton's good fortune that it remained for him to build the first viable steam vessel in 1807. The history of some of the steamboat components and early experiments reveal why this vehicle was launched seventeen years late.

The paddle wheel propelled the earliest steamboats. The ancient Romans before the Christian era carried their armies into Sicily on boats moved by paddle wheels turned by oxen. The Chinese used man-operated warships with paddle wheels about the seventh century. An old manuscript, dated 1430, in Munich, contains the sketch and description of a war vessel with four paddle wheels mounted on crankshafts and turned by four men.

The use of paddle wheels in 1450 to harness the current of a river is illustrated in a Latin manuscript preserved in Paris. The paddle wheels, turned by the force of the stream, serve to wind up a rope around their shaft. This rope, tethered upstream, pulls the boat forward.

Leonardo da Vinci, of course, produced several mechanical schemes for paddle-wheel propulsion. In one, dated about 1500, two pedals were used to drive a pair of paddle wheels, each with two blades. Leonardo also conducted research into the expansive properties of steam, but he never apparently experimented with steam as a motive power.

The earliest mention of steam and paddle wheels on a boat relates to 1543, but the paddle wheels were operated by manpower and the steam came from a caldron of hot water on the deck used for defensive purposes. Yet for a time the Spanish claimed Blasco de Garay, whose arrangement this was, as the inventor of the steamboat. This shows the confusion and hearsay that surrounds the whole subject of steamboats.

William Bourne, a British writer on military affairs, suggested another scheme for paddle-wheel propulsion in a proposal published in 1578:

> You may make a boate to goe without oares or sayle, by the placing of certain wheeles on the outside of the boate, in that sort, that the armes of the wheeles may goe into the water, and so turning the wheeles by some provision; and so the wheeles shall make the boate go.[1]

The "provision" was muscular power—either of animals or men—not steam.

The Ancients knew the force of steam. They had used the principle in the aeolipile of Heron of Alexandria about 50 A.D. Then the common people marveled as a round vessel rotated, seemingly by some miraculous force. Actually, tangentially escaping steam motivated it in an early example of jet propulsion. Philibert de L'Orme had proposed the use of an aeolipile to force smoke up a chimney in the mid-sixteenth century. Giovanni Branca offered more mature plans for a steam turbine at a somewhat later date. Salomon de Caus (1576–1635), a native of Normandy in the service of Louis XIII, described other applications of the aeolipile. He showed in an experiment that water could be driven up a tube to such a level as would balance the elastic force of the steam confined in the boiler. He used this principle in a steam-driven fountain. He suggested that the force could move vehicles on land and ships at sea and raise loads. For his genius, poor Salomon de Caus was imprisoned in a madhouse near Paris, but he is probably the first man to put on record the idea of the steamboat.

More specific (but still impractical) schemes for steam ship propulsion came from David Ramsey, a groom of the Privie Chamber in England, who in 1618 obtained a patent to "make boates runn upon the water as swifte in calmes, and more safe in stormes, than boates full sayled in greate wyndes."[2] In 1630, he obtained another patent for a steam engine, but there is no record that he ever tried one in a boat.

Edward Somerset, Marquis of Worcester, in 1663 published his description of "An Admirable and most Forcible Way to drive up Water by Fire" and obtained protection in that year by Act of Parliament for his "water commanding engine." [3] He invented the first practical steam engine, although there is no record that he ever applied it to a steamboat.

Nor did Thomas Savery who patented another steam engine in 1699, probably similar to Worcester's. He wrote a pamphlet in which he said, "I believe it may be made very useful to ships, but I dare not meddle with that matter." [4] From that we can infer that the British naval bureaucracy had pooh-poohed his idea. His "fountain engine" was a steam pump that could be used only in raising liquids. A large cylinder was filled with steam so that all the air was expelled. The steam was then cooled until it condensed or changed from vapor into water. Because it shrunk during this procedure and no more air was admitted, a vacuum resulted. This vacuum drew water upward into the cylinder through a long pipe. The fountain engine had gone through half its cycle. Next, valves closed the descending pipe, and opened another that extended upward. When fresh steam rushed into the cylinder, its pressure forced the water to rise through this second pipe out the top of the pump. Savery employed two cylinders that worked alternately. Although inefficient, it was better than anything that came before it and was used widely to pump water, especially in mines.

Denis Papin, a French engineer and Calvinist forced into exile by the revocation of the Edict of Nantes, may be the first who caused the steamboat to be an actual fact and not merely a figment of imagination. Papin designed the first engine with a piston. This device could make only a stroke or two at a time. When he applied the power of steam to boats, he reverted to Savery's more elementary form. In 1707, he actually constructed what may have been the first steamboat, which he ran on the River Fulda in Hanover. Some historians, including James T. Flexner, doubt that his vessel was propelled by steam. When the crunch came, Papin had to use muscle power. Whether propelled by steam or muscle, the Papin vessel didn't survive for more than a few short voyages.

Thomas Newcomen next deserves mention for his atmospheric or piston engine, which he patented in 1705 and built the first full-sized example of near Wolverhampton, England in 1712. Because the weight of the air was the motivating force of the engine, the top of the cylinder remained open. A beam, hung from its center like a seesaw, was attached at one end to the piston rod and at the other to the pump. It was weighted so heavily at the pump side that, when no other force intervened, the piston was pulled to the top of the cylinder. While the piston was rising, steam at normal atmospheric pressure ran into the closed cylinder beneath it. With a valve shut, this steam was condensed, creating a vacuum in the cylinder. The force of the air then pushed the piston down for the work stroke. When the piston was down, the valve opened, permitting new steam to enter as the weight on the beam pulled the piston up. Relying for power on steam's capacity to shrink when cooled, this engine wasted the other half of the cycle—water's capacity to expand when turned into steam.

Although the Newcomen engine proved the best means yet to pump water, it failed to propel boats. It was too heavy, too inefficient, and required too much fuel to work aboard a ship. Nevertheless, Jonathan Hulls tried it in 1736 on a towboat. He was laughed out of his native English village with a rhyme like this:

> Jonathan Hulls
> With his patent skulls
> Invented a machine
> To go against wind and stream;
> But he, being an ass,
> Couldn't bring it to pass,
> And so was ashamed to be seen.[5]

Others who tried the atmospheric engine on boats, all unsuccessfully, included Joseph Gautier in 1753 who experimented with it in Nancy, France, and the Swiss pastor, J. A. Genevois, who attempted it in Berne, Switzerland, in 1760.

In the competition which the Academie des Sciences of Paris opened on the best manner of mechanical propulsion, Daniel Bernoulli (1700–82) proposed in 1753 a scheme of jet propulsion. We have already seen how this intrigued Benjamin Franklin, but the state of mechanical arts could not make this system work.

Yet the man who really made the steamboat possible was James Watt. He was born in a small Scottish seaport, Greenock, the son of a prosperous artisan. Surviving a weak, delicate boyhood, he went to Glasgow and then to London to learn the trade of a mathematical instrument maker. In 1757, he became the instrument

maker at the University of Glasgow. In 1763, Watt received a
model of a Newcomen steam engine to repair. Although he fixed
it, its efficiency distressed his Scottish soul. For many months he
mulled over the problem, finally deciding that the trouble lay in
the fact that during each stroke the cylinder had to be heated to
the temperature of live steam, two hundred and twelve degrees,
and then cooled to produce condensation, about sixty degrees. Of
course, large amounts of heat were lost in this periodic change of
temperatures, while the first steam that entered the cold cylinder
condensed at that moment and was thus lost (besides putting
water where it was not wanted). In addition, the constant heating
and cooling took so much time that the engine could make only a
few strokes a minute. He decided that the remedy was to keep the
cylinder always above the boiling point. But how?

One Sunday afternoon in May, 1765 (six months before Robert
Fulton was born), Watt took a walk in the ancient Glasgow Green.
As he entered the gate at the foot of Charlotte Street and had just
passed the old wash house, a world-shaking revelation hit him.

"The idea came into my mind," Watt recalled, "that as steam
was an elastic body it would rush into a vacuum, and if a communication was made between the cylinder and an exhausted vessel, it would rush into it and might there be condensed without
cooling the cylinders . . . I had not walked further than the Gold-
house when the whole thing was arranged in my mind." [6]

He soon rigged a separate container, attached it to the cylinder
and created a vacuum in it into which the steam could be sucked
from the cylinder. The cylinder could then be kept hot, while the
separate container (condenser) remained cold. That was the basic
idea, but Watt also invented a pump, motivated by the engine and
attached to the condenser, which would preserve the vacuum by
sucking out the water created by condensation and any air that
might have leaked in.

Thus, the engine that would move Robert's steamboat was conceived about three months after he was.

Although Watt's model convinced him the idea would work,
more than three years passed before he took out a patent on it—
January 15, 1769. A poor man, he had a living to earn, and he
also had to solve various mechanical difficulties. Even with the
patent, he made little practical headway toward producing
engines for another five years. To make money, he surveyed canal

routes and did various other civil engineering chores. Finally, he met a rich manufacturer and capitalist, Matthew Boulton, who operated the Soho Engineering Works in Birmingham. The two formed a partnership, Boulton securing a patent extension for twenty-five years. Boulton left the inventing side to Watt, but handled the business end of Boulton, Watt and Company.

The first full-sized engine was not set up and put to work until March 8, 1776. Designed only to move pumps, the engines Boulton & Watt now began to construct in large quantity were not suitable for driving a vehicle on land or water. While Watt was a daring mechanical genius, he was a timid businessman. As long as the firm could sell everything they made, why go afield? His mechanical creativity, however, eventually forced him to do so. In 1781, he patented methods for achieving rotary motion with the steam engine, and in 1782, a double-acting engine. In listing possible applications, he mentioned a steam carriage, but not a boat. Nevertheless, he had created a device that could achieve man's age-old dream of moving against wind, current, and tide.

You would think that some enterprising Englishman would now have entered the lists with the first practical steamboat. A seafaring people, the British had the world's greatest navy and fleet of commercial vessels. The world's best steam engine was available in England. Yet, nothing happened. Their sea-going success with sail may have blinded them to the possibilities of steam.[7]

Actually, Frenchmen next contributed the steps toward steamboat development. The Comte d'Auxiron had made some abortive experiments about the same time that Watt was working on the engine. J. C. Perier in 1775 fitted up a small boat with steam which did cause it to move, yet it moved so little and so slowly that Perier gave up in disgust. The first documented real success in steamboat propulsion was achieved on July 15, 1783 on the river Saone near Lyons by the Marquis Claude de Jouffroy d'Abbons who used a double ratchet mechanism (similar to that of Papin) to get rotary motion for his paddle wheels. His rival Perier contrived that he be denied the support of the Academie des Sciences, and then the Revolution forced the aristocrat to flee the country.

Another steamboat pioneer in France, from about 1775, was the Abbé Etiènne d'Arnal, Canon of Alais Cathedral. In 1780 he retired from the church and devoted himself to mechanics and the steam propulsion of river boats. His first proposals were pub-

lished at Paris soon afterward, and, in 1782, he obtained a fifteen-
year concession for river steamboats. He had several plans, but
the Revolution postponed them. In 1799, he resumed his experiments on the river Rhône, but with no success.

The Englishman, Joseph Bramah, did obtain a patent in 1785
to propel a vessel by either a paddle wheel or what we would now
call a screw propeller, but which Bramah called a fly. No evidence
exists that Bramah ever built a steam vessel.

The scene next shifts to America and John Fitch and James
Rumsey. We have already seen the way in which Fulton's life
touched theirs. Fitch's experiments, from 1786 on, spurred Rumsey to launch a steamboat that operated twice for limited times
and distances on the Potomac River at Shepherdstown, Virginia in
1787. His boat traveled three miles per hour in August of that
year. The engine he used was roughly similar in principle to
Watt's early single-acting devices. To reduce the complication of
the apparatus, Rumsey placed his steam cylinder on top of another cylinder which functioned as a pump. The two cylinders
had a single piston rod, which passed through the division between them and carried the motion of the steam piston directly to
the plunger of the pump. When the steam piston rose, the
plunger descended, driving water through a pipe out of the stern
of the boat.

Although he used a separate condenser and an air pump, details of the apparatus show that Rumsey only partially understood
the advantages of Watt's discovery. He tried to keep his steam piston tight by shooting a steam of water on top of it at every stroke,
an expedient inferior to Watt's principle of keeping the steam cylinder always at the boiling point.

His brother-in-law, Joseph Barnes, who built the boat, tells us,
"The boat moved up the river against the current with about two
tons on board, besides the machinery, at 2 mph." But the boiler
opened at several joints, and the trial ended a half success.

Rumsey had the boiler repaired and set December 3, 1787 for
his next public trial. A local hero, General Horatio Gates, was
aboard and his colorful memoir, written years later and em-
bellished still more by others, is the usual source for the episode.
It is suspect: he claimed for instance, that the lady passengers
carried parasols—in December!

His certificate signed at the time simply says: The vessel moved "straight against the current at the rate of 3 mph." He believed that the machinery was very imperfect "and by no means capable of performing what it would do if completed." Yet he thought that the invention would become common "as the machine is simple, light and cheap, and will be exceedingly durable, and does not occupy a space in the boat of more than 4 feet by 2½." [8] The Reverend Robert Stubbs and other prominent citizens signed affidavits with the same wording.

On December 11, the boat voyaged again; this time the speed was given in liberally signed affidavits as four miles per hour, and the load as three tons. Six gentlemen added, "We think the machinery does not weigh more than six or seven hundred weight, and is not included in the burthen mentioned above." A local newspaper account, however, highlights a flaw in the motivating power: "If some of the pipes of the machine had not been ruptured by the freezing of water . . . which ruptures were only secured by rags tied round them, that the boat's way would have been at the rate of 7 or 8 mph." [9] That probably explains why Rumsey laid up his boat, never to move it again.

Rumsey's boat worked on the principle of simplified jet propulsion. It had much to commend it, as did Fitch's paddle wheel idea which at that time and state of technology proved more effective. So, Fitch really defeated Rumsey in the steamboat race, both from the point of time and of the practicalities of that day, but neither man won the rewards. Fate and their own vanities defeated them both.

The Delaware was about the poorest river Fitch could have chosen for his steamboat business venture because its wide reaches were ideal for sailboat operation and its wide banks ideal for fast coaches. From the middle of May until the end of September, 1790, Fitch's *Perseverance,* covering about eighty miles a day, had logged between 2,000 and 3,000 miles. It traveled at six to seven miles per hour, which was faster than the sailing packets, but not as swift as the stage coaches. Although its fare was only half that of the stage between Philadelphia and Burlington, N.J., it could not compete. Fear of boiler explosions deterred customers. At this time, a Fitch and Rumsey pamphlet war over their rival claims of precedence included accusations about the safety of each other's versions and increased the public's fears about the "floating sawmill."

If the Delaware was the worst body of water that Fitch could have chosen, the Potomac, Rumsey's choice, was a second poor selection. Its shoals and falls made operation of any boat in the late eighteenth century almost impossible except for relatively short stretches.

While the rivers chosen for the scenes of their experiments thwarted the two rivals' success, their temperaments cost them their last hope. When Fitch wasn't fighting with Rumsey, he was quarreling with his partner and financial backers. His scruffy appearance and tempestuous manners alienated many potential supporters,[10] notably Washington and Franklin who backed the attractive Rumsey instead. But when Fitch quarreled with his chief engineering aide, Harry Voight, he couldn't get the *Perseverance* to run properly. His financial supporters dropped him and he lost an important asset, a monopoly to operate steamboats in Virginia waters (which then also included what is now West Virginia and Kentucky). Furthermore, his Pennsylvania monopoly became clouded because he could obtain no clearcut patent from the United States government.

Although Rumsey's vessel clearly operated at a later date than Fitch's and could boast nothing like the mechanical sophistication of the Delaware steamboats, Rumsey had powerful friends, a persuasive personality, and unmitigated gall. He rounded up all manner of important people in Philadelphia and organized something he called the Rumsean Society whose main purpose was to give him financial and political support.

Its first tangible success was in the area of patents. Congress passed the first United States patent law on April 10, 1790, primarily to deal with steamboat claims. The first "Commissioners for the Promotion of Useful Arts" were Secretary of State Thomas Jefferson, an inventor himself, Secretary of War Henry Knox, and Attorney-General Edmund Randolph. As soon as the law went into force, Fitch, Rumsey, and John Stevens of Hoboken, New Jersey (about whom, more later) applied for patents. For months, Fitch bombarded the commissioners with petitions claiming that he, as the "first inventor," should get royalties on all steamboats, his own and others. His incoherent petitions probably harmed him. The commission set February 1, 1791 as the day to hear all steam inventors—the three steamboat claimants and Nathan Reed and Isaac Briggs who invented steam carriages. The commissioners stalled. They had much other business, and the conflicting claims required time to study. They finally "settled" the issue on August 26, 1791, by granting rights to each inventor with exactly the same wording on each application. To add to the insult, the actual patent described Rumsey's boat, not the superior Fitch or Stevens vessels. Jefferson, who knew and admired Rumsey, probably was responsible for this injustice. In effect, the "solution" threw the whole matter into the courts.

As a result, no one could get a clear patent on steamboats without lengthy and expensive court action. This precipitated a race to get state monopolies on the steamboat because such a recourse was the safest (although not foolproof) way an inventor could protect his rights. It also served to delay the advent of a practical steamboat for another sixteen years.

Fitch seemed to win one more chance when Aaron Vail, U.S. Consul at L'Orient, France, got a French patent on his boat (at the price of a lucrative share in the potential for Vail) and Fitch sailed for France in February, 1793. He arrived, however, in time for one of the bloodiest battles of the French Revolution which, surging around L'Orient, Nantes, and the only good foundry in the area, ruined his prospects. He waited there and in Britain until the next spring. When he returned to America, aged fifty-one, his productive life was over. He died, an alcoholic, in Bardstown, Kentucky in 1798.

Rumsey, in the meantime, suffered parallel ups and downs of fortune. The Rumsean Society raised $1,000 to finance a trip to England for him. He had many letters of introduction, including one to the firm of Boulton & Watt. On July 3, 1788, he visited the engine builders and he so impressed them that they virtually offered him the invention of the steamboat free. The deal was for Rumsey to patent the Watt engine in the United States on a profit-sharing basis. Also, the firm promised to back Rumsey in getting an English steamboat patent and to put their engine-building expertise at his disposal. Boulton & Watt evidently thought the matter settled and on July 14th instructed their lawyer to withdraw a previously-filed warning that they would oppose the efforts of anyone else to patent steamboats. Although, as we have seen, Rumsey's jet propulsion idea was not feasible in the technological development of that time, it is reasonable to assume that Watt's mechanical genius could have solved the difficulties. Yet he never had a chance to do so. Incredibly, Rumsey turned the deal down. The action so enraged Boulton and Watt that they refused ever to have anything directly to do with steamboats again.

Rumsey made his mistake—one of the major errors in business judgment of all time—because of bad advice from one Benjamin Vaughan, a London business man to whom he had another letter of introduction. Vaughan had advised Rumsey that he could evade Watt's patents by building the engines in Ireland. This kind of devious dealing apparently appealed to Rumsey. We will remember the earlier evidence that he lacked scruples in that he never corrected the impression that he had invented a steam boat when he had actually created only a *stream* boat.

Boulton put an end to the proposed partnership when he wrote Rumsey: "It appears to me there is so great change in your sentiments that I have little hope of our forming any connection, because you reason as if you expected shadows to counterbalance substance." [11]

Although Rumsey's failure to connect with Boulton & Watt can, on the surface, be attributed to over-sharpness, bad advice, and poor luck, the basic problem also lay in Rumsey's (and Fitch's) attitude toward invention. As experienced engineers and manufacturers, Boulton & Watt knew that a general idea requires much working out of details. Rumsey (and Fitch) thought they had invented the steamboat as soon as they conceived of it.

For the next two years, Rumsey connived and contrived to get money to build boats, all of which failed. The Rumsean Society dunned him for money as did his English backers. He had to waste much of his energy keeping out of debtors' prison. When he entered into a partnership, March 25, 1790, with Daniel Parker (originally from Watertown, Massachusetts) and Samuel Rogers, he thought his troubles over—until they went bankrupt. Almost destitute, he then won a reprieve with a job at £10 per day from the Earl of Carhamton, an Irish peer who was interested in building a canal. That didn't last long. Next he invented a mill superior to the previous mill inventions he had sold to the Rumsean Society. He raised money on the strength of it. Rogers' and Parker's fortunes improved; after more than a year away from his steamboat, Rumsey could return to it. In December, 1792, he tried it out, and it was moderately successful. With great reluctance, he agreed to a public trial.

Before that could take place, he had to attend a meeting of the committee on mechanics of the Society of Arts, where he intended to defend a device he had invented for equalizing the water on water wheels. He gave the learned members a lecture on hydrostatics which so impressed them that they invited him to write the resolution which they would adopt to praise him. He rushed to a table, began writing, then stopped in puzzlement. He tried to speak, but nothing emerged but a gabble of words. Friends rushed him to a doctor—too late. He died on December 18, 1792. Doctors still sometimes cite the incident as an example of apoplexy brought on by emotion.

The next year, Parker and Rogers tried out the vessel on the Thames. A London periodical reported that "a pump of 2 feet diameter, wrought by a steam engine, forces a quantity of water up through the keel. The valve is then shut by return of the stroke, which at the same time forces the water through a channel of pipe of about 6 inches square, lying above and parallel to the kelson out of the stern under the rudder which has a less dip than usual, to permit the exit of the water." [12] It achieved a speed of four knots, half that of Fitch's boat on the Delaware.

Significantly, Fulton and Rumsey had met in London. Possibly they were even good friends because they shared common traits. Fulton probably met Parker through Rumsey. The younger inventor and the Watertown, Massachusetts man would have business dealings.

Fulton wrote a detailed analysis of "Messrs Parker and Rumsey's experiment for moving boats." It's an impressive document, technically. He faults Rumsey because "the engine was not loaded to its full power, the water was lifted four times too high, and the tube by which the water escaped was more than five times too small." He believed that, although jet propulsion offered promise, "the power of the engine cannot be applied to advantage by this means." [13]

In June, 1793, a few months after the Parker-Rogers half-success with Rumsey's boat, Fulton for the first documented time tried his hand with the steamboat. Following some thought, he decided that the steamboat propulsion method should imitate that of the most effective sea animals, fish. Fish moved themselves by a springing motion of their tails. A steamboat's paddle should imitate the tail of a salmon. He experimented with a model and kept painstaking records of his results. Although the fish idea would prove unworkable, Fulton's experiments and record-keeping should have given other steamboaters pause. He was adding a new dimension to the art or science of inventing.

However, Robert put aside the steamboat for the time being because canals absorbed his attention during the English canal craze. Years later when he was trying to defend his steamboat monopoly in court, he would try to claim steamboat precedence as a result of his experiments in 1793. They are important, but not because of his claims. They started him thinking along new lines, though his thoughts wouldn't reach even partial fruition for another ten years—and not in England or America, but in France.

In the meantime, several others experimented with steamboats. For example, Serafino Serrati in 1787 tried out the jet propulsion principle, but no documentary evidence survives to support the claim of some Italian scholars that he built a full-scale steamboat which navigated on the river Arno about 1789.

Better substantiated are the experiments of Patrick Miller and William Symington. Miller was a new-style capitalist like Boulton—a breed then fairly common in Great Britain, although still rare in America. He had grown so wealthy from his Carron Iron Works in Edinburgh that he could retire at the age of fifty-four and buy an estate at Dalswinton where he followed all sorts of scientific pursuits, including boat propulsion. In 1786 and 1787, he and an assistant, J. Laurie, tried out double-hulled paddle vessels at Leith in Scotland. Thirty men straining at the capstans achieved a speed of 4.3 knots before they exhausted themselves. Miller looked around for some form of mechanical power. James Taylor, his sons' tutor, introduced him to William Symington, whom the English still nominate as the inventor of the steamboat.

Symington was the son of a millwright. He had been educated for the ministry, but machinery attracted him so much that he found a job at the Wanlockhead Lead Mine, kept free of water by one of Watt's engines. In daily contact with this marvel, he speculated on other uses for it. First, he toyed with steam carriages, but the state of the roads defeated him. He next turned to "working boats on canals," developing what he claimed was an improvement of Watt's engine—better methods of condensation, a rotary motion achieved with a "fly wheel to regulate and equalize the effects of the steam engine" by using "the alternate action of ratchet wheels."

He secured a patent on this on June 5, 1787, over Watt's protests of infringement. In 1788, Miller hired him to put his engine in a boat. He had done so by October of that year, and a trial was made on the 20th of that month. Symington reports, "in the presence of Mr. Miller and various other respectable persons . . . the boat was propelled in a manner that gave such satisfaction it was immediately determined to commence another experiment on a much extended scale." [14]

Taylor testified that the vessel went five miles per hour. For several weeks the steamboat chugged around a little ornamental lake on Miller's estate, before the engine was unshipped and mounted as a trophy in the wealthy man's library.

The relative speed with which Symington built his machinery shows the advantages Englishmen had over Americans in engineering matters. They had the craftsmen and equipment at hand to make the necessary devices. Furthermore, they could fairly easily get the financial backing that was much more difficult to find in America.

Miller next financed Symington in building a larger vessel which was ready by November, 1789—amazingly quickly by American standards. The two cylinders of the engine were upside down, the piston rod connected to the opposite ends of a beam suspended below them. A belt of rope leading from the ends of the beam looped over a wheel hung above and between the cylinders. Another belt transmitted the oscillating motion of the wheel to two paddle wheels.

Weaknesses in both the engine and the paddle wheels spoiled a trial in November, 1789. A second experiment came early in December after repairs. Testimony differs as to the speed achieved: Symington claimed five miles per hour; others said up to seven. At this point, inexplicable fate stepped in again to delay once more the advent of a working steamboat. Either Miller and Symington quarreled or something caused the money man such distress that he fired the other, despite the fact that he had built a vessel that had worked at least as well as any thus far, including Fitch's and Rumsey's.

Miller offered a partnership to Boulton & Watt, but they turned him down, disillusioned forever with the steamboat after their experience with Rumsey. Thus, anger in Miller, old age and conservatism in Boulton & Watt, naivete and greed in Rumsey, alcoholism in Fitch, and poor patent-law provisions in the United States all conspired to delay the steamboat.

Watt probably made the steamboat impossible in England because he refused to let his engine be used on any British boat. His firm challenged in the courts anyone who developed an engine remotely similar to theirs. For example, Boulton & Watt had Jonathan Hornblower, who developed the compound engine, sent to jail as an infringer for daring to apply his important discovery to the Watt engine. It is speculative—but likely—that the source of Miller's displeasure with Symington lay in a Boulton & Watt threat to sue for infringement. This theory would also explain why Symington didn't touch the steam engine or steamboats again until 1801, after Watt's basic patent had expired in 1800. Symington then found a new financial supporter in Thomas, Lord Dundas of Kerse, a governor of the Forth and Clyde Canal. He provided him with £7,000 to build a steam tugboat for his canal. Symington drew plans, patented them, and began work. He inclined the cylinder of his condensing engine at an angle of forty-five degrees and drove the paddle wheel at the stern with a crank and without the intrusion of any other machinery.

This excellent solution worked well in a March, 1803 trial when the tug pulled two barges loaded with seventy tons for nineteen and a half miles in six hours. But the other owners killed the project, claiming that the vessel would damage the canal's banks. The tugboat (named *Charlotte Dundas* after the Lord's daughter) was left to rot, and still another chance for the development of a practical steamboat rotted with her.

Years later, Symington claimed Fulton had ridden in the tug. This is unlikely because in 1803 Fulton was in France but he may have seen the vessel when she was laid up. When he was in England during his attempt to sell his submarine-torpedo system, he spent much of his spare time—of which he had a good deal—investigating every lead he could find about steam vessels. He undoubtedly at least familiarized himself with the Symington patents.

Fulton probably also read the patents taken out in 1788 by Robert Fourness and James Ashworth of Yorkshire. The two inventors in partnership are believed to have first tried a small steam-driven paddle boat on the river between Hull and Beverley, about 1787. As a result of these experiments, the partners constructed another steamboat, which was sent to London and assembled there. It had three steam cylinders which acted on a three-throw crankshaft; steam came from a copper boiler. The two paddle wheels on either side were each fitted with twenty fixed radial floats and could be raised or lowered as needed. This vessel ran trials on the Thames and underwent rigorous tests for the Prince Regent (later King George IV) who is said to have bought it later as a pleasure boat. The vessel burned, supposedly set afire by watermen who feared the introduction of steam power.

We know that Fulton knew the intimate details of Stanhope's *Ambinavigator.* The lord obtained patents in 1790 and conducted experiments with a model fitted with duck-foot paddles. An experimental flat-bottomed copper-sheathed vessel of two hundred tons without masts or sails, called the *Kent,* and based on his ideas, was started in October, 1792 by Marmaduke Stalkartt, an author on naval architecture, and launched in March, 1793. Trial runs on the Thames achieved a speed of three miles per hour. Stanhope and Fulton had copious correspondence on this project, the American recommending paddle wheels instead of the duck feet. Stanhope never followed the other's advice. The engine from the *Kent* was used later, in 1802, to drive the first steam dredger built for the British Admiralty. Stanhope had to use the Newcomen engine because Watt refused permission to use his engine.

Another steamboater about this time was John Smith of Lancashire. He too used the Newcomen atmospheric engines. Seven paddles on each side of the boat made eighteen strokes per minute. The vessel first moved in June, 1793 down the Sankey canal to Newton at a speed of about two miles per hour. Ridiculed in Liverpool for his efforts, Smith replied: "Before 20 years are over, you will see this river (the Mersey) covered with smoke." [15]

Fulton undoubtedly also studied the experiments in marine screw propulsion made about 1794 by William Lyttleton, a London merchant who had obtained a patent for what he called an "aquatic propeller." This device consisted of a triple-threaded screw of a length equal to one complete turn. It was supported in a frame which could be hung beneath the stern outside the rudder, or at the bow or sides. It was to be driven by men working at a winch connected by ropes and pulleys with the screw shaft;

Lyttleton mentioned that a small steam engine could be used. The trials disappointed the backers because manpower achieved speeds of less than two miles per hour. They deserve mention because of the presence of Colonel Mark Beaufoy who used these and other experiments to draw conclusions about water resistance. Fulton studied his work with great care and profit.

The American had other experiments to study, both in England and in France. For instance, Edward Shorter of London obtained a patent in 1800 for a "perpetual sculling machine." This device consisted of a screw with two or more blades "similar to the sails of a windmill," which was submerged in the vessel's wake, but supported by a buoy to prevent it from sinking too deeply. The screw was driven by a capstan, which was connected with a horizontal shaft above the water line. This shaft was hooked by a universal joint with a second shaft which was inclined downwards and carried the propeller at its outer end. By means of ropes attached near this outer end, the vessel could be steered. Although intended primarily for the propulsion by manpower of large vessels in calms, Shorter stated that his propeller could also be driven by a steam engine.

Meanwhile, the French were not idle. The first French patent for marine screw propulsion was obtained in 1803 by Charles Dallery of Amiens. He planned an experiment in the same year, but he couldn't raise enough money to proceed.

Detailed plans for a steamboat with side paddle chains were patented in 1802 by M. Desblancs, a watchmaker and mechanic of Ain, France, who had collaborated in the experiments of Jouffroy d'Abbans. This was a shallow, flat-bottomed craft, without sails and steered by a simple hand tiller. Although it didn't work in a trial on the river Saone, Desblancs gave Fulton a fright. The American had resumed work on the steamboat in France and for a time, he feared that the Frenchman would win success before he did.

10 | STEAMBOAT ON THE SEINE

No one ever embraced a project that would bring him fame and apparent fortune more reluctantly than did Fulton take up the steamboat. He had toyed with the vessel in 1793, but had dropped it for other things. Several reasons account for his attitude. The most significant probably was that so many people had tried it before him: If they had failed or only half-succeeded, why should he waste time and energy on a forlorn enterprise?

Another explanation has to do with his temperament. He could be coldly calculating on occasion, but if something gripped his emotions he would throw caution and logic to the winds. The submarine-torpedo system had so gripped him that it left room in his mind for little else in the mechanical field.

A third reason concerns money. Although Barlow, his chief and most consistent patron, was wealthy, Fulton could not expect him to underwrite more than one of his projects at a time. And his friend showed signs of restiveness even with the submarine-torpedo. Barlow fussed to his wife:

> Toot is calling for funds (for the submarine). Besides the 3,000 which I must pay him tomorrow, and 3,000 more at the end of the month, he wants 3,000 more still to build a new boat at Brest. I see no end to it; he is plunging deeper all the time, and if he don't succeed, I don't know what will become of him. I will do all I can for him, but the best way I can serve him is to keep a sheet anchor for him here at home that he might be sure to ride out a gale there if he can't keep the sea or get into port.[1]

Fulton did not dare ask him for much support for a steamboat which would probably be a project even more expensive than the submarine. Symington had needed £7,000 [2] from Dundas to build a steam tug.

If Chancellor Robert R. Livingston had not come on the stage as the United States minister to France, it's doubtful that Fulton would have returned to the steamboat. The new ambassador, who had assumed his duties in 1801, caught the inventor at the right psychological time. He had nearly lost hope of selling the submarine-torpedo either to France or England. Furthermore, Livingston had steamboat fever in its most virulent form and was looking for a partner to provide the mechanical expertise. He would put up the money. He passed his contagion on to his fellow-American, but Fulton could never muster the same passion for the steamboat that he had for the submarine.

Accounts differ about how the two met. One has it that Edward, the chancellor's youngest and favorite brother, visited the Panorama, met Fulton, learned of his inventive bent, and, in turn, introduced him to the ambassador. Another version indicates that Barlow, who made a point of knowing everybody worth knowing, introduced the two. Or it may have been that Fulton, who had also now cultivated the knack into a science of meeting and impressing important people, introduced himself. In any event, their meeting soon blossomed into a formal partnership, signed on October 10, 1802. In the first clause of the agreement, Livingston inserted, with some justification, that he as well as Fulton had "tried various mechanical combinations" and that the invention derived from each of them. The rest of the contract makes it clear that the principles that would be used came from Fulton.

In connecting with Robert R. Livingston, Fulton had lined himself up with an important person indeed. The ambassadorship to France was only the latest in a long series of public positions for the chancellor. He belonged to the second, third, and fourth Provincial Congresses of New York, was a delegate from New York to the Continental Congress in 1775–77, and again in 1779–80. He was also a member of the committee which drafted the Declaration of Independence. He couldn't sign that document because he was absent on another public task—attending a meeting of the fourth New York Provincial Congress which turned itself into a state representative body and adopted the first state constitution,

Portrait of Robert Fulton by Charles Willson Peale.
(Independence National Historical Park Collection)

Robert R. Livingston, Chancellor of New York, held a monopoly on steamboat navigation on the Hudson River. He financed Fulton's Clermont *and became his business partner. This portrait was painted by Gilbert Stuart. (Museum of the City of New York)*

1809 patent drawing of a boat for carrying passengers and merchandise (probably the Clermont). *The boat was to be 150 feet long with a displacement of 168 tons including 34 tons for merchandise. (American Society of Mechanical Engineers)*

1809 patent drawing for the assembled engine and boiler of the North River Steamboat of Clermont. (*American Society of Mechanical Engineers*)

1809 patent drawing of machinery of the North River Steamboat of Clermont. (*American Society of Mechanical Engineers*)

Nineteenth century Russian watercolor of deck life on the Paragon *(one of Fulton's steamboats) by Pavel Petrovich Svinin. (The Metropolitan Museum of Art, Rogers Fund, 1942.)*

It was while apprenticed to a Philadelphia jeweler that the teenaged Fulton found his first marketable skill as a miniaturist. These miniatures are of Mr. and Mrs. John Wilkes Kittera. (Pennsylvania Historical Society)

Fulton painted this portrait of his greatest friend and patron, Joel Barlow, in 1805. (Indianapolis Museum of Art. Gift of Mr. and Mrs. Eli Lilly, Sr.)

Many sneered at Fulton's idea of steamboats and canals as essential to the growth of industry and trade in America. Robert Havell's aquatint of 1839 (less than 25 years after Fulton's death) shows busy traffic on the Hudson. (I. N. Phelps Stokes Collection, Prints Division, The New York Public Library, Astor, Lenox and Tilden Foundations)

Drawing by Fulton of his captive mines. (Robert Fulton Papers, Manuscripts and Archives Division, The New York Public Library, Astor, Lenox and Tilden Foundations)

Berthaux's 1810 lithograph shows Fulton's steamboat passing West Point. The Clermont's *first public run in August 1807 was to Albany and back (at an average speed of 5 miles per hour) with a stopover at Livingston's estate of Clermont. (I. N. Phelps Stokes Collection, Prints Division, The New York Public Library, Astor, Lenox and Tilden Foundations)*

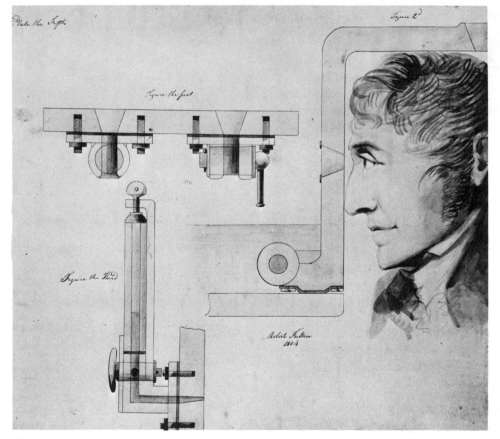

Fulton self-portrait in his submarine's conning tower. (Robert Fulton Papers, Manuscripts and Archives Division, The New York Public Library, Astor, Lenox and Tilden Foundations)

written by Livingston and others. He was instrumental in per-
suading the New York convention to ratify the new federal constitution. He became a Jeffersonian Republican, but lost the New
York State governorship to Federalist John Jay. President Jefferson awarded him the ambassadorship as a consolation prize. He
would prove able in that post, too, negotiating the Louisiana Purchase from Napoleon in 1803.

An engaging and affable man, he was the most popular ambassador from the United States since Franklin. At fifty-five years of
age when Fulton met him, he was a jowly, portly individual, but
his hair was still red. Beneath his genial exterior, he shrewdly
appraised his associates. Although Fulton impressed him, he saw
his faults, too, and kept a tight financial anchor on him during all
the years of their association.

Livingston was commonly called "chancellor" to distinguish him
from the other members in the large tribe of Livingstons. He had
been chancellor of New York State from 1777 to 1801, presiding
over the Court of Equity and attending the Council of Revision
which was empowered to veto legislation. In that position, he ad-
ministered the first oath of office to President Washington on
April 30, 1789.

Livingston was a member of a clan that owned the second largest of all the manors that the Dutch had originally parceled
out—120,000 to 160,000 acres granted in 1685 and reconfirmed
in 1705 by royal decree. The property started five miles south of
what is now the city of Hudson, extending twelve miles up the
Hudson River and from the river eastward to the state line of
New York and Massachusetts. At one point the manor was twenty
miles wide. Although divided and subdivided among the family
over the years, Robert's grandfather had 13,000 acres which he
called the Lower Manor of Clermont. (By the chancellor's time
the name was condensed to Clermont.)

In additon to his governmental and political activities, the chancellor also pursued scientific interests that related to his estate and
other business affairs, notably in farming and water transportation.

Every farmer in America at this time had trouble in moving his
crops to market. Transportation costs began assuming ever increasing importance, especially from the interior of the young nation. Moving a barrel of flour down the Susquehanna from the

Genessee Valley to Columbia and then to Philadelphia involved a carrying charge of $1.25, plus the likelihood of spoiling the flour on its slow journey in barges or behind plodding horses. To move a ton of freight from Pittsburgh to Philadelphia reached the impossible figure of $125; it cost $5 to freight one barrel of flour. It cost $1 a bushel to move salt one hundred miles or further, and the time could run into weeks. No wonder the cries for better transportation increased!

So Livingston came naturally by his interest in improved transportation. He also came by it through his talented brother-in-law, Colonel John Stevens who had achieved the rank as paymaster in the New Jersey militia during the Revolution. Stevens was born in New York City in 1749, just three years after Robert Livingston. When Robert married John's sister, Mary, two of the most socially prominent business families in America were united. The colonel owed his interest in steam power to John Fitch. While driving along the banks of the Delaware River near Burlington, in 1788, Stevens saw Fitch's steamboat pass up the river against the tide. He followed the boat to its landing where he got aboard and examined the engine and propelling paddles carefully. From that hour he became an unwearied experimenter in the application of steam power. Fortunately, he had the private means to afford to experiment. During the succeeding fifty-odd years Stevens lived in New York and on his estate across the Hudson in Hoboken. At the latter place he carried on all his experimental work with steam.

The best description of him is written by Stevens himself. Although he penned it in 1810 when applying for a European passport, he looked much the same for the previous and succeeding decades: "John Stevens of the City of New York in the State of New York. Was born in said City, 5 feet, 7 inches tall. Hair gray. Eyes gray. Complection rather fair. Forehead high and somewhat bald; person and visage neither thin nor very full. Age 51 years." [3] Ironically, for such a precise man he lopped ten years off his age. He was actually sixty-one at this time and had to write a hasty note making the correction. (He lived to be ninety.)

It will be recalled that in 1786 New Jersey had given Fitch a fourteen-year exclusive privilege to make and operate steamboats in the state. Stevens first petitioned the state legislature in 1788 for permission to place a steam engine on board a boat for experimental purposes. He then began intensive study of the whole subject of steam, devoting nearly two years to it. Stevens won his first ambiguous steamboat patent, along with Fitch and Rumsey, in 1791. He first tried a steam engine modified from the original Savery steam pump. During the succeeding six years he experimented with different propulsion methods, but like all other American steamboat aspirants he was hampered by the lack of men with mechanical ability and of tools.

Fortunately, he found Nicholas J. Roosevelt to help him. Roosevelt was probably the first man to engage in machine work in the United States. This great grand uncle of President Franklin D. Roosevelt came from a humble farming and hardware-merchant family in New York City and State. In 1794, he and two partners had bought six acres of land in Belleville, N.J. and built a foundry and machine shop there.

Stevens' absorption in steam infected Livingston who, in 1798, acquired Fitch's New York State rights to steam navigation. Livingston wanted to build a steamboat immediately and had more or less made up his mind to order a steam engine from Boulton & Watt. Stevens and Roosevelt, who had gone into partnership with him on the boat—but not on the New York State monopoly—dissuaded him from doing that and persuaded him to let Roosevelt build the engine at his shop in Belleville. Roosevelt at this time employed two Englishmen, John Smallwood and John Hewitt, both of whom had once worked for Boulton & Watt. With their knowledge and with the help of a German named Rohde who was skilled in making castings, Roosevelt believed he could build an engine for the Livingston-Stevens boat. And he did. The boat, sixty feet long, with a twenty-inch cylinder and two-foot stroke, made her trial in October 1798. It was unsuccessful, despite the fact that the engine worked well. The fault lay in the mechanism connecting with the paddles which had been suggested and insisted upon by Livingston.

Roosevelt found Livingston very trying, but, as a simple man, he could not resist the association with the patrician chancellor, despite the fact that his experienced employees put their objections in writing. They agreed to build "the chancellor's vessel according to his description, but if it should not succeed and have not the expected effect, do not blame us for it." [4]

When the experiment failed as predicted, the partners the next

year tried a new arrangement, this time from a Stevens plan for a
set of paddles in the stern with a crank motion, driving the boat
forward as they rose and fell. With Smallwood, Rohde, Hewitt
and Stevens aboard, the boat labored down the Passaic River from
Belleville to New York and back, but the mechanism shook the
vessel so badly that they abandoned this plan.

By this time, Livingston's ardor had cooled a little. He had
failed to meet the terms of the franchise granted him by New
York State concerning speed, but he persuaded the docile state
legislature to grant him an extension. He dropped out of the partnership when he went to France.

Stevens, however, was more determined than ever to solve the
difficulties. He designed and built a rotary steam engine to be
used with a screw propeller. He placed this combination in a little
twenty-five-foot boat in the summer of 1802 and used it occasionally in crossing the Hudson between New York and Hoboken. He wrote this about the little vessel: "She occasionally kept
going until cold weather stopped us. When the engine was in the
best of order, her velocity was about 4 miles an hour." [5] The
engine, while simple, was hard to keep steam-tight. That winter,
Stevens resorted again to the reciprocating engine.

In 1804, with the help of his seventeen-year-old son, Robert
Livingston Stevens, he launched and successfully navigated in
New York Harbor the first twin-screw-propellered steamboat. She
was named *Little Juliana,* and operated by a high-pressure, reciprocating steam engine and multitubular boiler, both of Stevens'
own design. Although small—scarcely more than twenty-five feet
in length—this boat was a marvel of engineering for that era.
Surely, Stevens would win the race to build the first commercial
steamboat!

Livingston was fully aware of what his brother-in-law was doing.
Why did he drop the partnership with him? He could have continued to deal with a relative who knew as much about steam
engines as anyone in America; who had designed and built four
different types—the steam pump after Savery, the Watt type with
separate condenser, a rotary, and a high-pressure, noncondensing
type. One explanation lies in a life-long rivalry between the two:
Each wanted to run the ship, and the captain had to be either
Stevens or Livingston. Ironically, Livingston would gradually relinquish the lead to Fulton in their partnership, but this did not
seem to bother him as much psychologically as taking second

place to Stevens would have. Another related reason was distance. Livingston wanted to be able to participate personally in any steamboat project. He could not do so when he was in France and Stevens in America. Finally, Stevens made his brother-in-law impatient. He insisted on experiments and still more experiments. As a dilettante in the field of invention, Livingston shared a misconception about it with Fitch and Rumsey. He thought that once the idea had come, the working out of details would follow quickly. He wanted fast action.

Fulton, we can be sure, promised it to him. A patent "for a new mechanical combination of a boat" was to be applied for in Fulton's name. The inventor was to go at once to England and there build a trial vessel for which he would borrow a steam engine. Livingston would put up £500 ($18,000). If the experiment failed, Fulton agreed to repay half of this in two years. If it succeeded, a passage boat was to be constructed under Fulton's direction at New York and was to steam to Albany. Livingston already feared Fulton's exuberant ideas and stipulated that the boat could not be more than one hundred and twenty feet long, eight feet wide, and fifteen feet deep. He added: "Such boat shall be calculated on the experiments already made with the view to run 8 miles an hour in stagnant water and carry at least 60 passengers allowing 200 pounds weight to each passenger." [6] The chancellor also insisted that he have the right to withdraw from the partnership if he wished after the initial £500 had been spent. Otherwise, the partnership would run for the same period as the United States patent and its extensions.

Fulton and Barlow worried that Livingston would meddle too much in the technical side of the enterprise. Barlow even suggested that Fulton bypass Livingston, go secretly to England, buy a boat, contract with William Chapman for a steam engine with a twelve-inch cylinder "and make experiments on that scale all quiet and quick." If the trial succeeded, Fulton could rush to New York, get an American patent and start operations. "I think I will find you the funds without any noise for the first operation in England, and if it promises well you will get as many funds and friends in America as you want." [7] Fulton and Barlow also considered approaching Parker, Rumsey's former partner. In the end, though, Fulton stayed with Livingston, and the fears of meddling proved unfounded.

Nevertheless, mutual distrust between the two persisted. Barlow

warned Fulton that Livingston thought the movement of the
engine might shake the boat to pieces. "I see his mind is not
settled . . . He thinks the scale you talk of going on is much too
large, and especially that part which respects money." On his side,
Fulton feared that Livingston would pirate his ideas. Barlow assured him: "I did not leave your memoir with Livingston above
three or four days. I brought it home for the same reason which
you suggest, and locked it up, and here it is. I talked with him yesterday again. He seems desirous of bringing the thing forward.
There is no danger of his trying to do the thing without you . . .
he sees too many difficulties in the way." [8]

The chancellor had perhaps become chastened after his mechanical suggestions had failed on the Stevens-Roosevelt boat. Livingston kept hands off the mechanical side of the enterprise and
applied his stronger talents for raising money and applying political influence. The New York State monopoly had expired for the
second time, but he again persuaded the legislators in Albany to
grant him and Fulton an exclusive right for twenty years, provided that, within two years, they ran a boat of twenty tons four
miles per hour on the Hudson. Some years later, Fulton wrote of
their collaboration, referring to himself in the third person: "To
produce the first useful steamboat, it required the fortunate circumstances of adequate genius and capital in the same person or
persons; he and Mr. Livingston had both." [9]

On January 24, 1803, Fulton made his first public steamboat
claim. He showed again his early skill in public relations by combining modesty, confidence, and clarity, with an impressive display of technical expertise. He would use this formula many
times, a major reason why the popular but erroneous impression
persists that he invented the steamboat. He sent to *Conservatoire
des Arts et Metiers* a description of a plan for towing linked barges
with a steamboat. This he intended "to put in practice upon the
long rivers of America," where there were few roads and labor
was expensive. In France which had good roads and cheap transportation, "I doubt very much if a steamboat, however perfect it
might be, can gain anything over horses for merchandise. But for
passengers it is possible to gain something on the score of speed."

"In these plans you will find nothing new," he disingenuously
wrote. Paddle wheels similar to those he used had been tried
before but failed. "After the experiments I have already made, I

am convinced that the fault has not been in the wheel but in the ignorance of proportions, speeds, powers, and probably mechanical combinations . . . Consequently, although the wheels are not a new application, yet if I combine them in such a way that a large proportion of the power of the engine acts to propel the boat in the same way as if the purchase was upon the ground, the combination will be better than anything that has been done up to the present, and it is in fact a new discovery." He promised to invite suitable sages to see his experiments. "If they succeed I reserve to myself the right of making a present of my labors to the Republic, or to reap from them the advantage which the law allows." [10]

He and Livingston had no intention of giving the steamboat to France or anyone else, but he often contrived to leave the impression that he was working for the public good.

Because Fulton still had no specific commitment from England on the submarine, he persuaded Livingston that they should make their first large-scale trial in France. They decided on a boat seventy feet long, eight feet wide and three feet deep, with paddle wheels twelve feet in diameter. Perier agreed to build the vessel and lease them an eight horsepower steam engine.

Fulton's inexperience with steam engines shows in his plan to use a pressure of thirty-two atmospheres. He designed an advanced modification of the pipe boiler. He injected just enough water into a red-hot chamber to produce the steam needed. This early example of the "flash boiler" could not hold up under the high temperature. Fulton returned to low pressure and to an ordinary boiler.

Desblancs chose this time to complain in the French journals that the American was infringing on his patents.

Earlier during the summer of 1802, Fulton had gone with Ruth Barlow to the fashionable spa of Plombieres. There he had tried out a model steamboat built for him (but Barlow paid for it) by the celebrated instrument maker, Etienne Calla. It was four feet long and two feet wide and driven by two strong clockwork springs. In a damned-up pond sixty-six feet long, Fulton tested the speed achieved by the model when the force of the clockwork was applied to the water by a variety of devices—paddles, a screw propeller, sculls, and wheels. These experiments convinced him that the most effective method of moving a boat was to attach flat, boardlike paddles to an endless chain that ran over pulleys at bow and stern. The chain would drive the paddles through the water for the work stroke. They were carried back through the air in an arrangement similar to the first that Fitch had tried. That's not surprising. Fulton had access to Fitch's plans as a result of Fitch's French patents.

As the inventor was congratulating himself, consternating news reached him from Joel Barlow. His friend had gone to the National Depot of Machines, a part of the patent office, "and there I saw a strange thing; it was no less than your very steamboat, in all its parts and principles, in a very elegant model. It contains your wheel oars precisely as you have placed them, except that it has four wheels on each side to guide round the endless chains instead of two." Barlow added, "I shall say nothing to Livingston of this model." [11]

Fulton rushed back to Paris, studied Desblanc's model, and decided that his rival posed no threat. Fulton was correct in this judgment because the Desblancs boat never did work; nevertheless, he decided to change his own plans from the paddles to paddle wheels to make absolutely sure that he would not infringe on the other patent.

Now the exasperating Frenchman dared plague him again. Fulton replied characteristically. First, he explained in a reasonable way how his plans differed from those of Desblancs. Next he stated quietly that his plans were superior to the other's. Then he offered Desblancs the "sleeves from his vest." He proposed that the two pool the rights to their inventions in France (where Fulton never intended to operate anyway) if Desblancs would pay half the cost of his experiments. Of course, the Frenchman did not agree.

And Fulton made another shrewd move. He pointed out that Jouffroy, who had returned from exile, actually had invented the steamboat twenty years earlier. Jouffroy rose to the bait and proceeded to debate Desblancs over who had precedence. The spotlight turned away from Fulton who quietly proceeded with his plans.

At this stage, Fulton freely admitted that many experimenters had preceded him with the steamboat. His and other evidence clearly proves that he was familiar with the work of nearly all his predecessors. Livingston had brought him up to date on all the steamboat experiments in America—not only those of Stevens and

Roosevelt, but also on the work of Oliver Evans and Samuel Morey.

Unlike Fitch, Rumsey and Stevens, Evans' imagination centered more on steam carriages than on boats. He never had the money to build a carriage, but he did become America's first manufacturer of practical steam engines, introducing high-pressure types which by their lightness and cheapness were ideally suited to the needs of the simple colonial industries. He had been apprenticed to a wheelwright and wagonmaker. As early as 1772, when only seventeen years old, he had begun thinking of possible ways to move wagons by means other than animal power.

For the next decade, Evans lived a hand-to-mouth existence on various mechanical jobs. His big chance came in 1782 when his two older brothers, who were millers, asked him to build a new flour mill for them. It took him nearly five years, but he ended with an automated conception. One man could handle such a mill, compared with the four needed in the old-fashioned mills. Yet, almost no millers would buy the "rattle trap." Although unfortunate for Evans, this disappointment forced him to turn to other things, including the steam carriage. Maryland granted him a monopoly in the state for it, stating that "it would doubtless do no good, but certainly could do no harm." [12]

Evans established the Mars Works in Philadelphia, announcing himself as an iron founder and steam engineer. He directed it until his unexpected death in 1819. He designed a small stationary steam engine. One was ordered in 1801 to drive a steamboat on the Mississippi. It never was installed, but was used to drive a sawmill instead. He also would build, in 1805, the first steam-operated vehicle in the United States, a dredger designed for the Philadelphia Board of Health which he named *Oruktor Amphibolos*, or amphibious digger. Before launching the scow, he put wheels under it, and by means of a series of belts between the engines and the wheels, transported it under its own power to the river. He drove it on land for a few days, inviting visitors to board it for twenty-five cents each. When he died at least fifty of his engines steamed away in many eastern states, but all were on land except for that one in the dredger.

Livingston probably met Morey in New York City in 1796 when he, his brother Edward, and Stevens took a ride in Morey's steamboat, steaming according to Morey's memoir "from the ferry as far as Greenwich and back, and they expressed very great satisfaction at her performance and with the engine." [13] Morey also claimed later that he "succeeded in making a boat run 8 miles an hour. He (Livingston) offered me at the time, for what I had done, $7,000 for the patent right on the North River (Hudson) and to Amboy. But I did not deem that sufficient and no bargain was made." [14]

Morey first experimented with a steamboat at his home in Orford, New Hampshire, on the upper Connecticut River in 1790 or a year or two later (he couldn't remember). He tested it by profaning the Sabbath, a good time for secrecy because almost everyone was in church. A prolific inventor, Morey sporadically returned to the steamboat between other projects. He probably abandoned it completely after about 1797 when he could find no financial backing. In that year at Bordentown, New Jersey he build a sidewheeler with two sidewheels mounted on one connecting shaft and driven by a crank in the middle; this was a vital and permanent development.

Morey was convinced that Livingston had cheated him. After steamboats had become common and lucrative, Morey allegedly exclaimed in Orford, "Damn their stomachs, those cusses stole my invention!" [15]

A mechanical genius, Morey took out twenty-one patents, among them an internal combustion engine, with carburetor, which the automobile pioneer, Charles F. Duryea, claimed was the father of all modern gasoline motors. In 1820 Morey used this in a small boat, *Aunt Sally,* which could have been the world's first motor boat. *Aunt Sally* supposedly gave him such scant satisfaction that he sank her.

Another American experiment may have escaped Livingston's and Fulton's attention—at least no evidence survives that they knew about Elijah Ormsbee. Early in the 1790's this carpenter and millwright from Cranston, Rhode Island, teamed up with David Wilkinson, a blacksmith and tinkerer from Pawtucket, to build a steamboat. First they put it in a dugout canoe. It didn't work. Then they borrowed a long boat from a captain of a large sailing vessel. Ormsbee attached the single-acting engine to a contraption that simulated ducks' feet. He steamed at about three miles per hour from Cranston to Providence and then steamed upriver to Pautucket to show Wilkinson. When they had to return the borrowed items, Wilkinson wrote, "Our frolic was over."

So, still another early steamboat trial failed or ended as a "frolic." Why? Because Americans, especially, had little experience in the techniques of invention and engineering. Many recognized the need for a steamboat, but their attitude toward mechanical wonders remained that of someone going to a sideshow.

Fulton's attitude toward inventing the steamboat differed importantly from that of all his predecessors. None before him had paid much attention to what *his* predecessors had accomplished. Each treated his experiments largely as an isolated phenomenon. Fulton, on the other hand, considered his as an historical development. After he had signed the partnership agreement with Livingston, he turned to Boulton & Watt for the steam engine. He had no intention of reinventing it, as other steamboaters had. Under the proviso that they had to get a permit to export it to America, he ordered a twenty-four horsepower, double-acting engine with a four-foot stroke, complete with air pump and condenser.

He carefully avoided disclosing his purpose. "The situation for which the engine is designed and the machinery which is to be combined with it will not allow the placing of the condenser under the cylinder as usual," [16] he explained ambiguously. He enclosed sketches of how he wanted the parts arranged, but he advised the firm to use their "better judgment" and hold to his specifications only if they would do so "without diminishing the power of the machine." Unlike other inventors, he at this stage left to the experts the final decisions. For example, he asked them many questions: What size should the boiler be? How should it be designed and placed? Could he burn wood rather than coal? If so, would this entail design changes? What should the cylinder dimensions be and how fast would the piston travel? Because "in the place where the engine is to work the water is a little brackish," he asked about the effect of a little salt in the condensing water.

Characteristically, Fulton had canvassed other engine possibilities before he had turned to Boulton & Watt. He had asked his old friend, Cartwright, in a letter on March 10, 1802, how he was faring with his alcohol steam engine. When the Englishman replied vaguely, Fulton wrote him: "It is with great pleasure I have received your flattering acount of your steam engine; and although attachment to you makes me believe everything you say, yet such belief is merely a work of faith, for I cannot see the reason why you have 13½ pounds purchase to the square inch.

. . ." [17] In other words, Fulton the engineeer wanted demonstrably proved facts, not guesses.

He had a relatively easy time with the steamboat, compared with his two major previous schemes, the canal and the submarine-torpedo systems. With canals, he proposed what would have resulted in a vast social revolution. With the submarine-torpedo, he had to fight the prejudices of the ages and the fact that technology had not yet sufficiently advanced.

With the steamboat, however, technology was more than ready. Even more important, the public was ready. Thousands of Americans had seen the Fitch, Rumsey, Stevens and Morey vessels. In addition, the Fitch-Rumsey pamphlet war on the steamboat had prepared others who had not see one. Finally, Americans particularly needed a steamboat to solve their transportation problems.

As the workers in Perier's shop began putting a vessel together on the Seine in front of their little factory, Fulton's always ebullient spirits must have soared. At last he would have a winner!

On July 24, 1803, he wrote gaily to his English friend, Fulner Skipwith, who had recently become a father:

> You have experienced all the anxiety of a fond father on a child's coming into the world. So have I. The little cherub, now plump as a partridge, advances to the perfection of her nature and each day presents some new charm. I wish mine may do the same. Some weeks hence, when you will be sitting in one corner of the room and Mrs. Skipwirth in the other, learning the little creature to walk, the first unsteady steps will scarcely balance the tottering frame; but you will have the pleasing perspective of seeing it grow to a steady walk and then to dancing. I wish mine may do the same. My boy, who is all bones and corners just like his daddy, and whose birth has given me much uneasiness or rather anxiety—is just learning to walk, and I hope in time he will be an active runner." [18]

According to Fulton's calculations, he would achieve a speed of sixteen miles per hour. He based this largely on the Beaufoy tables on water resistance. So confident was he that he publicized the expected speed. This interested Napoleon, who put aside his aversion to the inventor that lingered after the submarine-torpedo fiasco at Brest, and ordered representatives to witness the trial.

They almost saw nothing. Late in July, Fulton was awakened in the night at 50 Rue Vaugirard. A violent storm had sprung up.

His nearly completed boat had sunk! Imagine the thoughts running through his mind as he raced to the Quai Chaillot. Just as he was on the brink of the first important success in his thirty-eight years, fate would snatch it away from him again. He wrote afterward that he felt an anguish that he had never before known.

The despair gave him a demonic energy. Working for twenty-four hours without stopping even to eat, he and the workmen raised the boat and saved all the apparatus. Finally, he stumbled home, more ill than he had been even while an apprentice in Philadelphia. His physician, a Dr. Hosack, said his "laborious exertions" while "very much agitated" had brought on a "debilitation of the stomach" causing "chronic dyspepsia." In other words, his nerves failed, and the mental tension caused the stomach disorders. He would never quite free himself from these nerve-induced illnesses again.

Fulton rebuilt his steamboat. The *Journal des Débats* reported that on August 9

> a trial was made of a new invention, the complete and brilliant success of which should have important consequences for the commerce and internal navigation of France. During the past two or three months there has been seen at the end of the Quai Chaillot a boat of strange appearance, equipped with two large wheels mounted on an axle like a cart, while behind these wheels was a kind of large stove with a pipe as if there some sort of a small fire engine was intended to operate the wheels of the boat. Several weeks ago some evil-minded persons threw the structure down. The builder, having repaired the damage, received the day before yesterday a most flattering reward for his labors and talent.
>
> At six o'clock in the evening, helped by only three persons, he put the boat in motion with two other boats in tow behind it, and for an hour and a half he afforded the strange spectacle of a boat moved by wheels like a cart, these wheels being provided with paddles or flat plates, and being moved by a fire engine.
>
> As we followed it along the quay, the speed against the current of the Seine seemed to be about that of a rapid pedestrian, that is about 2,400 toises (2.9 miles) an hour; while going down stream it was more rapid. It ascended and descended four times from Les Bons Hommes as far as the Chaillot engine; it was maneuvered with facility, turned to the right and left, came to anchor, started again, and passed by the swimming school.

One of the boats took to the quay a number of savants and representatives of the Institute, among whom were Citizens Bossut, Car-
not, Prony, Perier, Volney, etc. Doubtless they will make a report
that will give this discovery all the celebrity it deserves; for this
mechanism applied to our rivers, the Seine, the Loire, and the
Rhone, would bring the most advantageous consequences to our internal navigation. The tows of barges which now require four
months to come from Nantes to Paris would arrive promptly in
from ten to fifteen days. The author of this brilliant invention is M.
Fulton, an American and a celebrated engineer.[19]

The article praised too lavishly. The vessel's top downstream
speed of about four miles per hour didn't approach Fulton's prediction of sixteen. Nor would it meet the terms of the New York
State monopoly, and Livingston was sour. The French savants did
not report favorably to Napoleon because of the speed deficiency,
and the newly crowned emperor is reported to have said about
the inventor: "Many charlatans or imposters . . . have no other
end but to make money. That American is one of them. Don't
speak to me about him any further." [20]

Yet Fulton worked hardest to reassure Livingston. He used the
failure to correct his calculations. The evidence showed that a
more powerful engine would easily achieve the four miles per
hour necessary. He made good use, we may be certain, of the
Journal des Débats' attribution of the first sinking accident to "evil-
minded persons." He may even have planted the explanation in
the author's mind. A less dramatic reason for the sinking is that
he had made the same mistake several of his predecessors had
committed: he put too heavy an engine in too flimsy a craft. He
would not make the same error when he built a new ship in
America for Hudson River traffic.

Before returning to the United States, however, he received
word from the British to come to England to demonstrate the
submarine. He always dropped everything when the slightest
flicker of interest developed concerning "the plunging boat." But
he reassured Livingston that he had to go to England anyhow to
supervise details about the Boulton & Watt engine; furthermore,
he would be gone only a few weeks. Actually, he would be in England more than two years and would not return to the United
States until late in 1806.

11 | TRIUMPH IN AMERICA

Fulton arrived in New York City on December 13, 1806, after an absence from America of nearly twenty years. He had left as a young miniature painter, he returned as a forty-one-year-old man looking, in the words of one of his workmen, like "an English nobleman . . . with his rattan cane in his hand." [1]

James Renwick, who had met Fulton about this time and who wrote his biography published in 1845, describes him this way: "Fulton was in person considerably above the middle height; his countenance bore marks of intelligence and talent. Natural refinement and long intercourse with the most polished societies both of Europe and America had given him grace and elegance of manners. His great success . . . never for a moment rendered him arrogant or assuming. Fond of society, he was the soul of the intelligent circle in which he moved. . . ." [2]

Livingston, who had left public life in 1804 and had been chafing ever since for Fulton's return to America, wrote Fulton a letter of introduction to his brother-in-law, John Stevens, because he wanted his protege to pick up the latest technical intelligence and to begin at once on the steamboat: "The principal object I have in view is to introduce to your acquaintance Mr. Fulton, with whom you will be much pleased. He is a man of science and embued with the best of principles . . . He is the inventor of a diving boat which is extremely original. . . ." [3]

It's not known if the two met at this time, but they did later and regarded each other warily. Stevens, in fact, would become one of the few people who knew Fulton personally and disliked him . . . but not all the time. Such was Fulton's charm that they would

133

have an on-again, off-again relationship and even would become partners for a while.

Fulton did not plunge full-time into steamboat affairs immediately. Livingston was annoyed, but the inventor went to Washington, D.C., anyhow, to visit the Barlows who had recently moved into a sumptuous home which Joel named Kalorama ("Beautiful View"). It no longer exists, but it stood at the top of what is now 22nd Street, about three hundred yards east of Rock Creek.[4]

Originally, Fulton was supposed to be a joint owner with the Barlows, but his steamboat ventures would tie him to New York City. He called it "our Athenian Garden in America" and on his first visit designed a summer house for "the grounds of our mansion." Whenever he visited Washington, that summer house became his home. He made many other suggestions for improvements over the years, both structural and in furnishings. He proposed a piazza across the rear. Thomas Barlow, Joel Barlow's nephew, worried—unnecessarily as it turned out—that his uncle would leave Kalorama to Fulton upon his death.

Barlow dammed up Rock Creek for a grist mill and to permit the inventor to conduct his submarine and steamboat experiments. Interest in the submarine-torpedo system had revived, thanks to Barlow's lobbying and Britain's increasingly vigorous attitude toward American ships doing a thriving trade with France. The "king" of Kalorama had money and friendship still invested in Fulton's project. The inventor had hardly been back in the United States a month before proposing a series of experiments with torpedoes to the government, undeterred by the fact that two foreign powers had already rejected them—including England who was now the enemy. He gave a demonstration in January at Kalorama, witnessed by James Madison, Secretary of State, and Robert Smith, Secretary of the Navy. The show impressed them sufficiently to make funds available for a demonstration in New York harbor. This took place on July 20, 1807, but established nothing new. It repeated his English demonstration.

Fulton describes the experiment:

> The brig was anchored, the Torpedoes prepared and put into the water . . .; the tide then drove them under the brig near her keel, but in consequence of the locks turning downwards the powder fell

out of the pans and they both missed fire. This discovery of an error in the manner of fixing the locks to a torpedo has been corrected. On the second attempt the Torpedo missed the brig; the explosion took place about one hundred yards from her and threw up a column of water 10 feet diameter, sixty or seventy feet high. On the third attempt she was blown up.[5]

On the day after this experiment, Fulton wrote an open letter to the governor of New York State and officials of New York City. Consider its effectiveness in the light of events the previous month. The British frigate *Leopard* overhauled the United States frigate *Chesapeake* within sight of the Virginia coast and demanded the right to search her for deserters. When the commander of the unprepared *Chesapeake* refused, the *Leopard* fired three broadsides that killed three Americans and wounded eighteen others. The crippled *Chesapeake* submitted to the seizure of four deserters and then limped to Norfolk. This humiliation so aroused the country that war seemed imminent.

> Having now clearly demonstrated the great effect of explosion under water, it is very easy to conceive that by organization and practice the application of the torpedoes will, like every other art, progress in perfection. Little difficulties and errors will occur in the commencement, as has been the case in all new inventions: but where there is little expense, so little risk, and so much to be gained, it is worthy of consideration whether this system should not have a fair trial. Gunpowder within the last three hundred years has totally changed the art of war, and all my reflections had led me to believe that this application of it will in a few years put a stop to maritime wars, give that liberty on the seas which has been long and anxiously desired by every good man, and secure to America that liberty of commerce, tranquility, and independence, which will enable her citizens to apply their mental and corporeal faculties to useful and humane pursuits, to the improvement of our country and the happiness of the whole people.[6]

While Livingston fretted over poor progress with the steamboat, Fulton pursued what the chancellor considered as still another distraction; he helped Barlow put to press the handsome edition of *The Columbiad.*

Fortunately for the chancellor's peace of mind, Fulton did turn down one other project. The Secretary of War, General Henry

Dearborn, had asked him to advise the government on canals. He refused in this letter of March 20, 1807:

> I am infinitely obliged by the proposal of the President [Jefferson] that I should examine the ground and report on a canal to unite the waters of the Mississippi and Lake Pontchartrain, and am sorry I cannot undertake a work so interesting and honourable. The reason is I now have ship builders, blacksmiths and carpenters occupied at New York in building and executing the machinery of my Steam Boat, and I must return to that City in ten days to direct the work 'till finished, which will probably require 4 months. The enterprise is of much importance to me individually and I hope will be of great use in facilitating the navigation of some of our long rivers. Like every enthusiast, I have no doubt of success. I therefore work with ardor, and when adjusting the parts of the machine I cannot leave the men for a day. I am also preparing the engines for an experiment of blowing up a vessel in the harbour of New York this Spring. The machines for this purpose are in great forwardness and I hope to convince the rational part of the inhabitants of our cities that vessels of war shall never enter our harbours or approach our Coasts but by our consent.[7]

If the *Chesapeake* affair had led to war, Fulton would have turned from both canals and steamboats to devote all his time to the cherished submarine-torpedo system. But the crisis blew over, and so did hopes for the torpedo. He almost reluctantly returned to the steamboat.

"I will not admit that it (the steamboat) is half so important as the torpedo system of defense and attack," he wrote Barlow, "for out of this will grow the liberty of the seas, an object of infinite importance to the welfare of America and every civilized country." [8]

Fulton's preoccupation with other matters did contribute to delays in getting the steamboat into the water. Because the proprietors could not have it afloat by April 1807, the deadline under terms of the state monopoly, Livingston had to return to the legislature for a third extension. This time he outdid himself, winning even more favorable terms than before. Originally, the monopoly was to run from 1803 to 1823, provided a ship was operating by April, 1807. The legislature of New York now passed a law to prolong the exclusive privilege of Livingston and Fulton by five years for each additional boat they should establish, provided that the whole time should not exceed thirty years.

In the meantime, the Boulton & Watt steam engine had languished in the United States Customs warehouse in New York since November, 1806. Some writers have speculated that neither Livingston nor Fulton could afford the duties to claim it—it's more likely they wished to avoid paying storage fees themselves on it and waited until April to take possession. Although rich, Livingston hated to spend money. Like most wealthy Americans of that era, he had most of his resources in land, not liquid assets, and sometimes had trouble raising cash. He often complained of Fulton's "grand ideas," meaning expensive notions.

Fulton gave him progress reports now and then. On July 4, for example, he reported to Livingston, "I have all the wheels up; they move admirably."

Soon an unexpected problem arose—vandalism. The boatmen on the Hudson could see the strange contraption taking shape. Fulton had to hire guards. On June 7 he noted in his expense book, "$4.00 to the men for guarding the boat two nights and a day after the vessel ran against her." [9] Through June, July and August numerous such entries appear.

Total expenses were exceeding budget. Livingston refused to put up another dollar. Fulton probably didn't have a spare dollar. The partners tried to borrow the needed $1,000, Livingston turning to his brother-in-law, Stevens. The colonel offered criticism, not money: The screw propeller was much better than paddles; the vessel was so long and thin that it would founder. Stevens was wrong on the first criticism, given the state of the technology at the time; he was correct on the second. In 1909 when a Hudson-Fulton celebration attempted to duplicate the boat, the builders had to make her wider than the original because of the danger of capsizing. That the vessel did not turn over in 1807 remains a miracle.

The partners went to other friends. "Never did a single word of encouragement or bright hope or a warm wish cross my path," Fulton complained. "Silence itself was but politeness, veiling doubts or hiding reproaches." [10]

After Fulton had spent an entire evening in a fruitless attempt to get one friend to invest, he returned the next morning, this time asking for any amount the friend could spare. The friend lent him $100, stipulating that his name never be mentioned so that he could avoid ridicule for a connection with "Fulton's Folly," as the venture had become known. On the same basis, the inventor persuaded nine others to invest.

Building the wooden boat itself proved the least of Fulton's worries. Charles Brown [11] built the hull at Corlears Hook on the East River. When completed, Fulton had it towed to Paulus Hook Ferry where he had acquired land for a workshop.[12]

At the shop, Fulton built the framing and gearing for the engine, and the paddle wheels. He experienced trouble with all this work because he had only blacksmiths and carpenters to do the unfamiliar operations. He complained also of the brittle copper produced by Revere in the Boston area and turned to a local producer, Harmon Hendricks, buying $653.12 worth from him. Hendricks was probably one of the ten friends who lent him $100 anonymously. Later, he did not object to making public the fact that he had bought twenty shares in Fulton and Livingston's North River Steamboat Company which they would soon organize.

When the partners launched the vessel they did not christen her as simply calling her "The Steamboat" was distinctive enough because she was the only one in the world at the time. Fulton never called this first boat the *Clermont.* When he began commercial voyages, he advertised her as *The North River Steamboat* (North River is an alternate name for the lower Hudson River). When he rebuilt her in the winter of 1807–08, the reconstruction was so extensive that he had to re-register her and named her in his application *The North River Steamboat of Clermont.* In all subsequent documents, he called her *The North River Steamboat* or simply *The North River*, yet, somehow, *Clermint* stuck as the name. The *Hudson Bee* for May 13, 1810, reported: *"The North River Steamboat* is believed to be the first one built on the river and has lately been known by the name *Clermont,* that is in books."

Fulton obviously tacked on *Clermont* as an afterthought. Could adding the name of his partner's estate have been a sop to Livingston, who never was delighted with many aspects of his steamboat adventures? [13]

Certainly, the derision over the project and the unseemly scramble for money did not please the dignified Livingston. Furthermore, the boat looked strange—more like the section of a raft. It was one hundred and fifty feet by thirteen, its flat deck rising only a few feet from the water. It had not a single curve because Fulton's experiments had convinced him to cut off the bow and the stern at an angle of sixty degrees to form a point. A small mast for sails that rose at the stern contrasted with a fat chimney at midships. The steam engine remained in plain view because it was not decked over. Most startling were two clumsy paddle wheels on each side, with no guards.

Fulton adopted the awkward-looking design as a result of calculations by Mark Beaufoy. He was the first to apply them practically. Beaufoy showed that the important factors in the total resistance of a solid were skin friction, proportional to the wetted area and to the square of the velocity; bow and stern resistance, proportional to the square of the sine of the angle of obliquity of the bow and stern.

Although these conclusions have been modified today, they are still partially correct. Henry W. Dickinson, of the British Science Museum and author of a Fulton biography that appeared in 1913, comments, "It is hardly too much to say that he (Fulton) was the first to apply theoretical investigations to practical ship design, so entirely was the latter at that time a question of rule of thumb." [14]

On August 9, 1807, four years to the day from the time Fulton had tried out his Seine steamboat, he made an unheralded run on the Lower Hudson, despite the fact that the paddle wheels were not completed. Even so, the boat achieved three miles an hour, proving, the inventor wrote the chancellor, that "she will, when in complete order, run up to my full calculations. I beat all the sloops that were endeavoring to stem the tide with the slight breeze which they had; had I hoisted my sails, I consequently should have had all their means added to my own. Whatever may be the fate of steam-boats on the Hudson, everything is completely proved for the Mississippi, and the object is immense." [15]

This Fulton steamboat had auxiliary sails; indeed, several more would have them. It would take a while before the inventor would have complete faith in his engines and nearly a year would elapse before he would have full confidence in their potentials for the Hudson. Livingston, who had negotiated the Louisiana Purchase, may have fired Fulton with his Mississippi enthusiasm. Many Americans at this time had caught the first whiffs of "Western Fever." It's not surprising that the ebullient Fulton had it, too.

Eight days after the almost secret trial, Fulton was ready for a full-blown trip. On August 17, 1807, a hot summer day, about forty stylish passengers arrived, mostly members of the Livingston clan or their hangers-on. The ladies' bright dresses and bonnets and the men's ruffles and lace were not designed to withstand the soot and sparks from the dry pine wood used to keep the engine throbbing. The furnace gave off an almost continuous plume of dark smoke, adding a galaxy of sparks whenever the fire was stirred. The vessel looked like a sawmill mounted on a raft and set on fire. The passengers' and the inventor's correct clothes were soon smoke-stained and peppered with tiny burns. But Fulton's serious, almost sad, expression and lordly manner tempered the festive atmosphere in which the voyage had begun. Let Fulton describe the event:

> The moment arrived in which the word was to be given for the boat to move. My friends were in groups on the deck. There was anxiety mixed with fear among them. They were silent, sad, and weary. I read in their looks nothing but disaster, and almost repented of my efforts. The signal was given and the boat moved on a short distance and then stopped and became immovable. To the silence of the preceding moment now succeeded murmurs of discontent, and agitations, and whispers and shrugs. I could hear distinctly repeated—'I told you it was so; it is a foolish scheme: I wish we were well out of it.'
>
> I elevated myself upon a platform and addressed the assembly. I stated that I knew not what was the matter, but if they would be quiet and indulge me for half an hour, I would either go on or abandon the voyage for that time. This short respite was conceded without objection. I went below and examined the machinery, and discovered that the cause was a slight maladjustment of some of the work. In a short time it was obviated. The boat was again put in motion. She continued to move on. All were still incredulous. None seemed willing to trust the evidence of their own senses. We left the fair city of New York.[16]

Engineer Fulton omitted one romantic note in his account. At the start, Chancellor Livingston made a little speech which, almost as a by-the-way, included the announcement of the inventor's betrothal to the young Livingston cousin, Harriet. Although she was a famous beauty of the day, daughter of the aristocratic Walter Livingston and his wife, Cornelia Schuyler (who also boasted impeccable lineage), it's doubtful that the former Pennsylvania farm boy wasted much time then congratulating himself on such a catch. He had to succeed with this voyage, or the marriage might not materialize. The Livingstons rarely allowed their daughters to marry failures.

The chancellor, at least, seemed to have few doubts. After the betrothal footnote, he went on to say that the "name of the inventor would descend to posterity as a benefactor to the world," and that before the close of the century vessels might even go all the way to Europe without sail. John R. Livingston was then heard to say in an aside to his cousin John Swift Livingston: "Bob has had many a bee in his bonnet before now, but this steam folly will prove the worst yet!" [17]

Besides presenting a strange sight along the Hudson, the steamboat offered unusual sounds. The paddle wheels creaked and splashed, the engine pounded, and the Livingston passengers contributed near-hysterical gabble after the first suspense was over. They even gathered at the stern to sing *Ye Banks and Braes o' Bonny Doon:*

> Ye banks and braes 'o bonny Doon,
> How can ye bloom sae fresh and fair;
> How can ye chant, ye little birds,
> And I sae weary fu' of care? [18]

Some boatmen on the river, recognizing a business rival, shook their fists at the unfair competitor. Now and then, a more knowledgeable landowner, recognizing the steamboat as a venture of the great patroon, would get into his own sail boat and move out to offer congratulations. Miraculously, none got entangled with the paddle wheel or otherwise collided with her.[19]

On the all-night trip the ladies retired to an impromptu cabin at the stern. Most of the men stayed up. Amid the racket and rocking of the narrow vessel, no one could have slept much. If they lacked sleep, they did not want for food, supplied by the black cook, Richard Wilson, and served by stewards both black and white.

As dawn broke, more people lined the banks because news about the steamboat traveled faster than its laborious five miles per hour. Even so, it had met the terms of the New York monopoly and was going faster than most sail boats could navigate. Although the sloops made an occasional passage in sixteen hours, normally they needed four days to sail the one hundred and fifty miles from New York to Albany.

At Poughkeepsie, spectators agitatedly argued whether the day of judgment had arrived. A farmer allegedly gave the strange boat one look, then raced down the road to his house where he banged the shutters closed and locked the door until the apparition had passed.

The North River had left New York on Monday at one o'clock in the afternoon and arrived at Clermont twenty-four hours later, having traveled one hundred and ten miles at an average rate of less than five miles per hour. Upon their stop the second night at Chancellor Livingston's estate, Andrew Brink, captain of the vessel, rowed across the river and brought back his wife from their home to fulfill his promise to "take her to Albany on a boat driven by a tea kettle."

Fulton and his party resumed the voyage the next morning at nine A.M., arriving at Albany, forty miles up river, at five in the afternoon. They made the Clermont-Albany run at an average speed of five miles per hour.

The chief engineer on the up-river voyage was a Scot, perhaps the first of a long line of Scottish chief engineers. Unfortunately, he had a failing that fable, at least, attributes to Scottish engineers. He got so drunk on the layover in Albany that Fulton discharged him and promoted Charles Dyke, the assistant engineer, to the chief position. Fulton kept him for the rest of Dyke's life. Regrettably, that was not long because Dyke died in a steam machinery accident on a ferry to Brooklyn the inventor would establish in 1814. He was one early victim among many in the pioneering steamboat days. At Dyke's death, Fulton "was perfectly unmanned and wept like a child." He was quoted as saying, "I will give all I am worth to save the life of that man." [20]

On Thursday at nine in the morning, the *North River* left Albany and she arrived at the Chancellor's at six in the evening. The vessel resumed the voyage to New York at seven the following morning and arrived at the home port at four in the afternoon. In thirty hours, she had traveled one hundred and fifty miles at an average speed of five miles per hour. During the whole trip, coming and going, the wind was ahead of Fulton so that he could not use his sails. His steam engine drove him every mile of the way.

Only one newspaper of the twenty in New York, *The American Citizen*, noted the event at the time:

> Mr. Fulton's Ingenious Steam Boat, invented with a view to the navigation of the Mississippi from New Orleans upward, sails today from the North River . . . to Albany. The velocity of the steamboat is calculated at four miles an hour. It is said it will make a progress of two against the current of the Mississippi, and if so it will certainly be a very valuable acquisition to the commerce of the Western States.[21]

Livingston and Fulton may have given the reporter the speculation about the Mississippi as deliberate misdirection to the rival sail boaters who showed increasing alarm at the pending competition. Although the partners had Mississippi on their minds, too, the Hudson remained upper-most. They had a commercial monopoly for that river, but nothing similar for the western one.

Fulton's correspondence reveals that he felt a mixture of elation and letdown at the success of the first trip. On the one hand, he could write to Barlow in this triumphant vein:

> I overtook many sloops and schooners beating to the windward and parted with them as if they had been at anchor. The power of propelling boats by steam is now fully proved. The morning I left New York, there were not perhaps thirty persons in the city who believed that the boat would ever move one mile an hour, or be of the least utility, and while we were putting off from the wharf, which was crowded with spectators, I heard a number of sarcastic remarks. This is the way ignorant men compliment what they call philosophers and projectors. Having employed much time, money, and zeal in accomplishing this work, it gives me, as it will you, great pleasure to see it fully answer my expectations. It will give a cheap and quick conveyance to the merchandize on the Mississippi, Missouri, and other great rivers, which are now laying open their treasures to the enterprise of our countrymen; and although the prospect of personal emolument has been some inducement to me, yet I feel infinitely more pleasure in reflecting on the immense advantage that my country will derive from the invention.[22]

In another mood, Fulton could also write this about the maiden voyage to Albany and back: "When all seemed achieved, I was the victim of disappointment. Imagination superseded the influence of fact. It was then doubted if it could be done again, or if done, it was doubted if it could be made of any great value." [23]

After her return from her historic voyage, Fulton laid up the steamboat for more than two weeks to complete her equipment and carry out improvements. To remain as light as possible on the first trip, the inventor had not completed the cabins or decking. On August 29, he wrote the chancellor about his progress:

> I have been making every effort to get off on Monday morning, but there has been much work to do—boarding all the sides, decking over the boiler and works, finishing each cabin with twelve berths to make them comfortable, and strengthening many parts of the ironwork. So much to do and the rain which delays the caulkers will, I fear, not let me off till Wednesday morning. Then, however, the boat will be as complete as she can be made—all strong and in good order and the men well-organized and I hope nothing to do but to run her for six weeks or two months.[24]

The steamboat was not ready, however, to start till Friday, September 4, when she made her first commercial voyage, carrying fourteen paying passengers. Advertisements appeared in the *Albany Gazette* of September 2—the first public notice of the new method of transportation—and for the following three weeks: "The North River Steam Boat will leave Pauler's Hook Ferry on Friday the 4th of September at 6 in the morning and Arrive at Albany at 6 in the afternoon (of the next day). Provisions, good births [sic!] and accommodations are provided." [25]

No births were recorded on the boat, but she had twenty-four berths: twelve at the bow for men and the same number at the stern for women. The steamboat's accommodations may have surpassed her sailing rivals, but they were not luxurious by today's standards.

A passenger on the steamboat's first commercial return trip from Albany to New York was M. Michaux, a botanist sent by the French government to survey forest potentials in the New World. The translation of his account of the trip:

> At this time decked sailing vessels arrived and departed every day with twenty-five or thirty passengers. The passage generally took 36 or 48 hours, according as the wind or tide was more or less favorable.

We had been three days at Albany when the arrival from New York of a vessel propelled by steam was announced. This boat, which was decked . . . was commanded by the inventor, Mr. Robert Fulton. Many of the inhabitants of the city and strangers who were there at the time went to visit it. Every one made his remarks upon the advantages consequent upon the new means of navigation, but also upon the serious accidents which might result from the explosion of the boiler. The vessel was lying alongside the wharf: a placard announced its return to New York for the next day but one, the 20th of August, and that it would take passengers at the same price as the sailing vessels—three dollars.

So great was the fear of the explosion of the boiler that no one, except my companion and myself, dared to take passage in it for New York. We quitted Albany on the 20th of August in the presence of a great number of spectators. Chancellor Livingston, whom we supposed to be one of the promoters of this new way of navigating rivers, was the only stranger with us: he quitted the boat in the afternoon to go to his country residence which was upon the left bank of the river. From every point on the river whence the boat, announced by the smoke of its chimney, could be seen, we saw the inhabitants collect; they waved their handkerchiefs and hurrahed for Fulton, whose passage they had probably noticed as he ascended the river.

We arrived the next day between one and two o'clock at New York . . . The day after our departure from Albany, and a few minutes after Chancellor Livingston had acquitted us, Mr. Fulton expressed his surprise that, notwithstanding the number of persons who were going to New York, only two Frenchmen had the courage to embark with him.[26]

A Quaker who steamed on the *North River* recalled that a friend asked him, "John, will thee risk thy life in such a concern? I tell thee she is the most fearful wild fowl living, and thy father ought to restrain thee." John made the trip anyhow.[27]

Rumors also began to circulate that fumes from the steam engine would wither corn and apples and steal insidiously into the "birth rooms" to poison passengers. Nevertheless, the steamboat soon lured more and more people to try it. Fulton had to prepare a behavior code:

Regulations for the North River Steam Boat [28]
The rules are made for order and neatness in the boat, are not to

be abused. Judgment shall be according to the letter of the law. Gentlemen wishing well to so public and useful an establishment, will see the propriety of strict justice, and the impropriety of the least imposition on the purse or feelings of any individual.

Fulton also wrote instructions for Captain Brink:

As she is strongly man'd and every one except Jackson under your command, you must insist on each one doing his duty or turn him on shore and put another in his place. Everything must be kept in order, everything in place, and all parts of the Boat scoured and clean. It is not sufficient to tell men to do a thing, but stand over them and make them do it. One pair of Quick and good eyes is worth six pair of hands in a commander. If the Boat is dirty and out of order the fault shall be yours. Let no man be idle when there is the least thing to do, and make them move quick.

Run no risques of any kind when you meet or overtake vessels beating or crossing your way. Always run under their stern if there be the least doubt that you cannot clear their head by 50 yards or more. Give in the account of Receipts and expenses every week to the Chancellor.[29]

Except for some minor collisions with sailors, which may or may not have been accidental, all went well with the steamboat until November 13. On that date, just as she was leaving New York, "one of the axletrees broke off short and she was obliged to return." The axletrees were cast iron, so the break was not surprising. Fulton had the repairs made in one day, so sophisticated had his workmen become at Pauler's Hook.

Slowly, he had gathered a crew of technically competent employees, no mean feat in that period. He made his men feel that they were the elite, and they responded with almost uniform admiration for him. For example, Paul A. Sabbaton, his chief engineer, who saw his boss constantly for half a dozen years: "He had all the traits of a man with the gentleness of a child. I never heard him use ill words to any one of those employed under him no matter how strong the provocation might be,—and I do know there was enough of that at times; and ever and anon my mind recurs to the times when his labours were so severe. His habit was, cane in hand, to walk up and down for hours. I see him now in my mind's eye, with his white, loosely tied cravat, his waistcoat unbuttoned, his ruffles waving from side to side as his movements caused their movements; he, all the while in deep thought, scarcely noticing anything passing him." [30]

J. B. Calhoun of Brooklyn, in Fulton's employ for several years, described him as a tall, somewhat slender man, of fair, delicate complexion, of graceful, dignified bearing, and mild and gentle in his temper. Calhoun wrote: "His workmen were always pleased to see him about his shops." [31]

Another contemporary describes the inventor:

> "Among a thousand individuals you might readily point out Robert Fulton. He was conspicuous for his gentle, manly bearing and freed from embarrassment, for his extreme activity, his height, somewhat over six feet—his slender yet energetic form and well accommodated dress, for his full and curly dark brown hair, carelessly scattered over his forehead and falling around his neck. His complexion was fair, his forehead high, his eyes dark and penetrating and revolving in a capacious orbit of cavernous depths; his brow was thick and evinced strength and determination; his nose was long and prominent, his mouth and lips were beautifully proportioned, giving the impress of eloquent utterance. Trifles were not calculated to impede him or damp his perseverance.[32]

Helen Livingston, daughter of Gilbert R. Livingston and later the wife of William Mather Smith, described him when she was a young girl: "There were many distinguished and fine-looking men on board the *Clermont,* but my grandmother always described Robert Fulton as surpassing them all. 'That son of a Pennsylvania farmer,' she was wont to say, 'was really a prince among men.' He was as modest as he was great, and as handsome as he was modest. His eyes were glorious with love and genius." [33]

Even people not disposed to admire him, ended in grudging admiration. For example, Robert R. Thurston, head of Cornell's Department of Mechanical Engineering, who wrote a debunking biography of him published in 1891, said this about him:

> Nature had made him a gentleman, and bestowed upon him ease and gracefulness. He had too much good sense to exhibit affectation, and confidence in his own worth and talents gave him a pleasing deportment in all companies . . . He conversed with energy, fluency, and correctness; and owing more to experience and reflection than to books, he was often interesting in his originality.

In all his social relations he was kind, generous and affection-
ate . . .[34]

So, Fulton cast his spell over most contemporaries and even
over people who never knew him personally and had doubts
about his technical genius. But the inventor couldn't influence the
weather! By November 19, 1807, the Hudson froze from Albany
as far down as Coxsackie. Even so, the steamboat plowed through
the ice, although she did have to ride at anchor for seven hours. A
few days after this, Fulton decided to lay up the boat for the winter. He informed Livingston in a letter of November 20 about his
plans for the off-season:

> After all accidents and delays our boat has cleared 5 per cent on the
> capital expended, and as the people are not discouraged, but con-
> tinue to go in her at all risques and even increase in numbers, I
> think with you that one which should be complete would produce
> us from 8 to 10,000 dollars a year or perhaps more and that an-
> other boat which will cost 15,000 dollars will also produce us 10,000
> dollars a year; therefore, as this is the only method which I know of
> gaining 50 or 75 per cent, I am, on my part, determined not to
> dispose of any portion of my interest on the North River; but I will
> sell so much of my funds as will pay my part of rendering this boat
> complete and for establishing another, so that one will depart from
> Albany and one from New York every other day and carry all the
> passengers. It is now necessary to consider how to put our first boat
> in a complete state for 8 to 10 years—and when I reflect that the
> present one is so weak that she must have additional knees and tim-
> bers, new side timbers, deck beams and deck, new windows and
> cabins altered, that she perhaps must be sheathed, her boiler taken
> out and a new one put in, her axels forged and Iron work strength-
> ened. With all this work the saving of the present hull is of little
> consequence particularly as many of her Knees Bolts timbers and
> planks could enter into the construction of a new boat. My present
> opinion therefore is that we should built a new boat . . .[35]

The first steamboat cost $20,000, of which $3,000 had been
borrowed and $8,500 each supplied by Fulton and Livingston. As
usual, Fulton was optimistic. The profit on three months' operations had been $1,000. For the year 1808, he expected to earn
eight times as much, although the period during which the boat
would run would be only about three times as long.

During the winter of 1807–08, a new hull of increased beam was built. This explains why, in his patent specification in 1809, he sometimes gives the beam as thirteen feet, but in other places he records it as sixteen or eighteen feet. So changed was the original vessel, that it was virtually a new boat. With her, Fulton looked forward to 1808 as a banner year.

12 | FULTON THE GENTLEMAN

For a man who once had nothing in his pockets but his hands, Robert Fulton began 1808 on the crest of a financial, engineering and social success that would rival even a Horatio Alger fantasy. The son of a bankrupt, he seemed headed for certain wealth. Poorly educated until in his twenties, he was now widely acknowledged as America's leading engineer. Born of a modest family, he was marrying into one of the most celebrated clans in America, the Livingstons.

On January 7, the forty-two-year-old Robert wed Harriet, a handsome blond some twenty years his junior. We know little about her except that she was beautiful, played the harp—and second fiddle to her husband. He seldom alludes to her in his correspondence. Her references to him that survive often complain about money. One of Fulton's good paintings is of her mother, a woman much closer to him in age than his wife.

A miniature he painted of Harriet is so poor in quality that one can hardly believe that the same artist did both. The colors are pale, the result unattractive. It gives little hint of the beauty that numerous contemporaries attest to. When painting her, Fulton may have grown so self-conscious that an inhibited portrait resulted.

Like his father, Robert Jr. married a woman much younger than himself. Despite his claim that his paternal heritage did not count for much with him, Fulton at least followed his father's example to some extent in choosing a wife.

All the evidence indicates that he got along well with his new in-laws. Walter Livingston, his father-in-law, was the son of the Lord of the Manor, having inherited 28,000 acres of some of the best land in New York. On a height above the Hudson, he had built a palace that now became the Fultons' summer home. Their house in New York would be a distinguished address [1] with a magnificent view of the river and harbor. There he hung his works of art on the walls. The dining room with its handsome Sheraton furniture became the scene of pleasant hospitality. His china, a gift from Thomas Jefferson, was embossed with the coat of arms of the United States. He had servants and a carriage. The house no longer stands, replaced now by an office building.

Fulton's marriage almost cost him the Barlows' friendship. Ruth Barlow especially resented Harriet and made her views clear to Fulton. He naively thought that Ruth's pique would pass. Five months after the wedding he wrote "Ruthinda": "Shall we unite our fortunes to make Kalorama the centre of taste, beauty, love and dearest friendship, or by dividing interests never arrive at that comfort, elegance, or happiness for which our souls are formed?"

He said that he wanted his bride to study the example of "Ruthinda, dear Ruthinda, heart of love" so that she might "acquire all that is [in] her most endearing." [2]

No record has surfaced of how Harriet took this, but she probably reciprocated Ruth's resentment. When Fulton died, Ruth wrote her sister: "In him I have lost a sincere friend; in spite of her, he was such." [3]

Fulton smoothed over the estrangement during his lifetime, however. When he realized that the two families could not live together, he wrote a long, disarming letter to Barlow. "Believe me, my dear Barlow, your ease and happiness and that of the dear and amiable Ruthinda is as dear to me as my own." [4] He gave the reasons why he had to live in New York. He also totaled as best he could from memory his friend's capital invested in Kalorama to show that the Barlows could afford to live there alone. The letter, a model of conciliation, apparently satisfied Joel, if not Ruth.

Fulton's acute perceptions about people led him to put an end to Ruth Barlow's sulks. When business took him to Washington, he visited the Barlows without announcing his arrival. Later, Ruth wrote her sister that Fulton had "popped in upon us yesterday morning without the smallest notice. I was never more surprised.

You see, I had no opportunity to say no, and now I am glad he came this way." [5]

Ruth never accepted Harriet and put up with her only because she could no longer have Fulton alone. Even if Fulton had never married, however, he could not have lived in Kalorama because his many activities kept him away from Washington. Besides his increasingly pressing business affairs, art patronage took a growing share of his time. European noblemen had been collecting art for generations, but few private collections yet existed in the United States. Although his was modest, Fulton had one of the first in the new nation. He lent the canvases he had purchased in London to various academies. A show of these paintings sponsored by the Pennsylvania Academy of Fine Arts in Philadelphia caused entrance fees to soar to the unheard of total of $100 a month. He specified that the receipts be used to establish a school of drawing, an act that received suitable publicity.

Scientific and cultural societies lined up to elect him a member. Oddly, the torpedo probably contributed as much—or even more—to his technical reputation as the steamboat did. Senators and Congressmen knew about it and Barlow continued as his effective press agent.

And a trivial event made Fulton seem almost dazzling in his technical knowledge. A showman named Redheffer had exhibited to large crowds in Philadelphia a perpetual motion machine which he had invented. He then moved on to New York where he received wide publicity and attracted even larger crowds. One afternoon Fulton went to see it and the people who had gathered to marvel at the wheels spinning apparently forever.

Fulton watched. His trained eye noticed something familiar about the spin of the wheel. "This is a crank motion!" he said. The audience held back the outraged Redheffer while Fulton knocked away a partition, to reveal a cat-gut string running from a pulley through a hole in the wall. He traced the string to a back loft where he revealed "a poor wretch with an immense beard and all the appearances of having suffered a long imprisonment" turning a crank. With this incident, Fulton won as much notoriety as either the steamboat or the torpedo had achieved for him. The public at large now recognized him as a hero.

Fulton had chosen to headquarter himself in the most dynamic city in the United States. New York was wresting the cultural,

social and economic leadership of the new nation from Philadelphia. With 83,000 inhabitants in 1808, it had passed Philadelphia
to become the largest city in the United States. (When the first section of the Erie Canal was finished in 1819, it would win un-
disputed leadership. Fulton would play a small but significant role
in planning the canal.)

During the early months of the year, Fulton spent most of his
time superintending the finishing touches on the new *North River*.
As usual, he spent every penny of his own and of others whom he
could persuade to part with any. This time, he turned the vessel
"into a floating palace, gay with ornamental paintings, gilding,
and polished woods." He started the tradition of sumptuous
steamboats which flowered on the Mississippi.

He expanded from the two mean cabins on the old *North River*
to three with a total of fifty-four berths, a kitchen, larder, pantry,
bar, and steward's room. He soon found he had to enforce stricter
rules of conduct than on the first steamboat:

> It is not permitted for any persons to lie down in a berth with their
> boots or shoes on, under a penalty of a dollar and a half, and half a
> dollar for every half hour they may offend against this rule. A shelf
> has been added to each berth on which gentlemen will please put
> their boots, shoes, and clothes that the cabins may not be encum-
> bered . . . As the steamboat has been fitted up in an elegant style,
> order is necessary to keep it so; gentlemen will therefore please to
> observe cleanliness, and a reasonable attention not to injure the
> furniture; for this purpose no one must sit on a table under the
> penalty of half a dollar each time, and every breakage of tables,
> chairs, sofas, or windows, tearing of curtains, or injury of any kind
> must be paid for before leaving the boat.[6]

Although Fulton started the trend toward great elegance, the
accommodations in his day seemed unimpressive to many English,
including Fanny Trollope, novelist Anthony Trollope's mother
and a scathing commentator on the American scene. She wrote: "I
can hardly imagine any motive of convenience powerful enough
to induce me again to imprison myself in a . . . boat under ordinary circumstances." [7] In 1820, Charles Dickens wrote that he
went below deck on a steamer in the United States to seek his bed.
Instead, he found "suspended on either side of the cabin three
long tiers of hanging book shelves designed apparently for vol-

umes of the small octavo size. Looking with greater attention at these contrivances (wondering to find such literary preparations in such a place) I descried on each shelf a sort of microscopic sheet and blanket; then I began dimly to comprehend that the passengers were the library and that they were to be arranged edgewise on these shelves till morning."

The method of assigning berths was often by lots. Dickens saw "some of the passengers gathered around the master of the boat at one of the tables drawing lots with all the anxieties and passions of gamesters depicted in their countenances; while others with small pieces of cardboard in their hands were groping among the shelves in search of numbers corresponding with those they had drawn. As soon as any gentleman found his number he took possession of it immediately by undressing himself and crawling into bed. The rapidity with which an agitated gambler subsided into a snoring slumberer was one of the most singular effects I have ever witnessed." [8]

In June, more than a month behind schedule, the new *North River* was at last ready to steam. Or so it seemed. She made it to Albany, then broke down. The problem was not Fulton's engineering, but Livingston's. The patroon had left the mechanical side of the first *North River* venture pretty much to Fulton. Perhaps made foolhardy by the successes of 1807, the chancellor insisted that one of his own inventions, a wooden boiler, be used. He had tried it on stationary engines where it had worked with indifferent success. Fulton reluctantly agreed to use it on the steamer. In a wooden chest containing water, an iron furnace was placed, while metal flues carried the smoke several times through the water before it reached the chimney. The theory was that since wood conducts less heat than metal, its use would reduce fuel consumption. The fallacy lay in the fact that the wooden boiler would not stay tight for long.

Fulton replaced the offending device with a conventional metal boiler. It worked well, and the new ship proved more efficient than the old. The enrollment for the new boat was dated May 14, 1808 and stated:

> Robert R. Livingston of Clermont, Columbia County, State of New York, having taken and subscribed to the Oath required by the said Act and having sworn that he, together with Robert Fulton of the

City of New York, are citizens of the United States and sole owners
of the ship or vessel called the North River Steamboat of Clermont,
whereof Samuel Wiswall is at present Master, and as he hath sworn
he is a citizen of the United States and that the said ship or vessel
was built in the City of New York in the year 1807 as per enrollment 973 issued at this port on the 3rd day of September 1807 now
given up, the vessel being enlarged. And Peter A. Schenck, Sur-
veyor of the Port, having certified that the said ship or vessel has
one deck and two masts and that her length is 149 ft; 183 48/95
tons. That she is a square sterned boat, has square tuck; no quarter
galleries and no figurehead.[9]

Fulton would eventually build twenty-one steamboats, each with
some improvement over its immediate predecessor. In so doing,
he differed markedly from all the previous experimenters with
the first great invention that America offered the world. The new
nation could boast other new devices before this—the cotton gin,
for example—but none had the impact of the steamboat. The gin
required little that was new in the way of technology, although it
did transform agriculture in the South. The steamboat had a double impact. It required new and more significant technology to
build it, and it would transform the inland transportation system
of America.

For example, by 1813, Daniel French, a Pittsburgh inventor, had
built a steamboat adapted for the Mississippi, the *Comet,* and a
year later, the larger *Enterprise.* In May 1815 she steamed from
New Orleans to Louisville, the first steam voyage up the Mississippi and Ohio, although flooded river conditions detracted from
the accomplishment by making it easier. After Fulton's death,
Henry M. Shreve, an experienced keelboat captain, really ushered
in the Mississippi steamboat era. His *Washington* chugged under
normal river conditions in 1817 from New Orleans to Louisville in
twenty-one days, less than one-fourth the time normally required
by barges or keel boats.

Fulton and Livingston heirs fought Shreve in court because
they had a Louisiana monopoly that prevented him from steaming into New Orleans and on that portion of the river in the state.
The Fulton-Livingston interests would lose the legal battle, thus
permitting western steamboating to mushroom into one of the
marvels on earth.

Robert Fulton came late and reluctantly on the steamboat scene.

He did not invent the steamboat, but he built the first commercially practical vessel because he brought to it a novel approach—of a technologist, rather than of an inspired tinkerer. By the time he came along, there had already been enough gifted geniuses to build a score of steamboats that worked—but didn't last. He built vessels that survived.

Technology is the sum of the ways in which a society provides itself with the material things of civilization. Fulton was one of, if not *the* first of the technologists, a word that fails to appear in Webster's first dictionary of 1806. Fulton intuitively defined it, but called it invention: "Invention [technology] in mechanics consists in a new combination of the mechanical powers, or a new combination of parts of known machines producing thereby a new machine for a new purpose and thus . . . forming a machine which has in itself a new character or principle that is . . . performing something new and useful." [10]

He gathered material, devices, and objects together and made something new of them. Everything about the steamboat had been invented already. The principles of shipbuilding had been known for centuries. Watt had a well developed steam engine. The paddle was known, of course; so was the paddle wheel. The world had long known about various devices, such as pulleys and chains, to connect a source of power with an object to make use of the power.

Fulton brought all of these previously known devices together and welded them into a practical unit, in a manner that his predecessors had failed to do.

The early experimenters with the steamboat failed because of their attitude toward invention. They went at it in a perhaps inspired, but still slap-dash, way by trial and error. Often when they finally got something to work, they couldn't repeat the success because they had kept no records of their trials and errors. Not so with Fulton. He kept meticulous records. He built models before he ventured anything on a full scale. His long apprenticeship on the canals, with submarines, torpedoes, and his many other devices had taught him how to invent in a methodical manner—much more in common with the late nineteenth and twentieth centuries than with his own era.

In eighteenth century America, ingenious inventions were prized more for their "philosophic" interest than their usefulness.

Even though Benjamin Franklin was a craftsman and mechanic, this was his attitude. Inventions were considered oddities to be displayed in museums and sideshows. For instance, the Columbian Museum of Boston advertised in 1797 an automatic "canary bird which sings a variety of beautiful songs, minuets, marches, etc., as natural as life," and a clock that marked the hours by a mechanically motivated tableau showing "King Herod beheading John the Baptist, and his daughter holding a charger to receive the head." Charles Willson Peale, a Philadelphia painter, had great engineering abilities, but he used them to simulate sea battles on a miniature stage by the use of complicated devices.

Fulton had tried his hand at stagecraft with his Paris panorama, but he had found it unrewarding, both from economic and psychological standpoints. He wanted to make a lot of money, but he also wanted a dignified fame to accompany the acquisition of a fortune. The panorama qualified only marginally on the first count and not at all on the second. He made more money with his submarine and torpedoes, but it galled him that he had to carry on these activities secretly for the most part. Indeed, much of the fame that came to him from this source proved more negative than positive because many people thought his warfare methods barbarous.

Besides the attributes of training and attitude, Fulton had still another characteristic that put him in a unique position to build the first practical steamboat. By 1806, he had become half-European after two decades in England and France, but he remained an American, too. As a half-European, he had learned the scientific approach to studying all aspects of a subject. For example, James Watt spent infinite time and experimentation determining the elasticity of steam so as to then design the steam engine to take advantage of the findings. Some of the early steamboat experimeters didn't care a continental about the elasticity of steam. Even when their inspired experiments did happen upon a combination that accidentally made good use of this property, they didn't know why this had happened. Accordingly, they thought design and invention were much less complex than they really are.

Fulton did realize the importance of elasticity of steam. He did recognize the complexity of the steam engine. Thus, he had no intention of reinventing it, as other steamboaters had. He hit upon the astoundingly novel—and practical—course of buying the device from the world's acknowledged experts in the field, Boulton & Watt.

Yet his Americanism helped him, too. He knew in his bones the American farmer's need for better transportation to get his crops to market. He had the American instinct for tinkering and mechanics. Writing at the close of the eighteenth century, that famous French observer of America, M.G.J. de Crevecoeur, said: "The farmer and the artisan have more to do than they can perform; scarcity of men makes labor very dear; to supply the want of labor and time, the American is forced to invent, to think out new ways of augmenting his efficiency." [11] Fulton also had the American optimism of that era. Anything was possible.

Thus, he combined qualities hitherto separate. He read books and knew theory. Some of his predecessors had acted without great thought. Others had thought without much action. He did both. He had also profited from the best influences in the three nations—England, France, and America—that were leading or were about to lead the scientific and industrial revolution. And he had one more advantage. He had greater interest in results than originality. At least two of his predecessors, John Fitch and James Rumsey spent as much energy in disputing who had invented the steamboat as they did on the vessel itself.

Fulton, finally, added another contribution to the practical steamboat. He made a start at rationalized manufacture. Today, it is hard to imagine the state manufacturing was in during the early years of the nineteenth century in America. Many landowners, such as Jefferson whom Fulton admired as a result of Barlow's influence, actually feared what manufacturing would do to their agricultural paradise. "Manufacture must," the plantation owner wrote in *Notes on Virginia*, "be resorted to of necessity (as in Europe, where lands are scarce), not of choice. But we have an immensity of land . . . Those who labor in the earth are the chosen people of God . . . While we have land to labor then, let us never wish to see our citizens occupied at a workbench or twirling a distaff." [12]

The few skilled craftsmen living in the young nation had mostly immigrated from England and continental Europe. The lure of land even enticed many of them from their trade. Fulton's own father is an early example. As a result, manufacturing entrepreneurs in the first decade of the nineteenth century often had to make do with the dregs of society to man their shops. So hard pressed was Eli Whitney for artisans that he worked around the problem, inventing the concept of interchangeable parts in 1798 because he couldn't find skilled gunsmiths. By developing molds to make every part exactly like all the others, he found that boys off the farms could make guns. Yet neither he nor any other manufacturer could solve the lack of reliable materials. They were so crude that boiler joints, for example, blew out a dozen times a day, developing fatal leaks at the very moment when a workable head of steam had been raised.

This is why higher pressure steam boilers were in very low repute at this period, even though the principle was well understood. Bad accidents came so frequently that Watt had established limiting figures of two and a half to three pounds to the square inch. As late as 1838, the first all-steam trans-Atlantic ship, *Great Western*, ran on the pressures decreed by the sainted Watt. Colonel John Steven's *Little Juliana*, the world's first successful high-pressure-boiler steamboat, operated on 100 pounds of pressure. But she was scarcely more than a rowboat in size, and her boiler blew up finally, fortunately not hurting his sons, Robert and John Cox Stevens.

Sophisticated equipment, even by the standards of those days, could rarely be found in the United States. As we have seen, Oliver Evans couldn't sell many of his automated milling systems. A nine-day wonder was the machine to produce cold-cut nails, invented by Jeremiah Wilkinson in 1777 and improved by Jacob Perkins in 1790. Even though the Perkins design could turn out 100 times the output of a good manual laborer, hand-made nails still outnumbered machine-made by 1808.

An explanation lay in the chronic shortage of capital to buy or build equipment. Rare were the Americans who could afford to purchase machines from England, the source of most good apparatus in those days. From memory, Samuel Slater reproduced English spinning machinery, utilizing water power, for his mill in Rhode Island. The enormous sum of $250,000 was needed to build a crude iron mill in America in 1800. It was unthinkable for anyone in the new nation to finance a manufacturing facility on the scale of a Boulton & Watt.

Yet, worst of all, almost nobody approached manufacturing sys-

tematically. This resulted because the principles of mercantil-
ism—buy here, trade there, sell in still another place—carried
over into manufacture. One reason was that many of the early
manufacturers started out as merchants. Harmon Hendricks,
from whom Fulton bought copper for at least nine of his steamboats, is an example, beginning as an importer of metal. In addition to importing, he decided to make the metal himself and
bought a defunct copper mine and mill in New Jersey, thus trans-
forming himself into a manufacturer. Yet he also continued as a
merchant, calling himself a merchant investor until his death.

To build his first steamboats, Fulton had to go one place for the
hull, another for the paddle wheels, purchase his steam engine
from England, buy his copper from Hendricks, and hunt here,
there and everywhere for the hundreds of other parts and devices
required. Gradually he began to make more and more for his own
needs at his expanding facilities at Paulus Hook Ferry. Because he
had to, he systematized his own manufacture.

He brought to manufacturing the mental habits of an engineer.
So rare was engineering skill in America in those days that even as
late as 1823 we find the editor of a magazine called *The Portfolio*
gravely listing the qualifications of a canal engineer:

> The engineer should be well acquainted with the principles and
> use of the several instruments in geodesic operations . . . He
> should be qualified to survey with the utmost accuracy, embracing
> not only horizontal but vertical sections and lines, and to level with
> minute precision, making accurate allowance for the earth's curva-
> ture . . . In prosecuting the duties of his profession, he should be
> so familiar with the several mechanical powers, as to be able to
> apply them to the best advantage, as he will have a constant variety
> of occasions for resorting to their aid.[13]

As we shall see, if Fulton had remained content to be a builder
of steamships, instead of both a builder and operator, he probably
would have made an enduring fortune. Instead, he chose to be
primarily an operator and built steamboats only because nobody
else existed in the country to make them to his design and specifications.

Although an innovative man in many ways, Fulton grew in-
tensely conservative socially. A gentleman—and he now had
achieved that status—did not make things in the early 1800s

unless circumstances forced him to do so, as the gentlemanly Jefferson had advised. A gentleman engaged in the professions or politics, or managed estates or practiced in a few of the genteel trades such as banking, imports-exports, or shipowning.

Like a religious convert who adopts the tenets of his new creed with far more passion than anyone born into the faith, Fulton fast became the ultra-gentleman. The Pennsylvania farmer and tailor's son wrote to a person inquiring about his services as a shipbuilder: "I shall always hope that gentlemen contracting with me for steamboats . . . will not think or deal with me as a common ship carpenter or boat builder. Any gentleman who is disposed to do so, shall have their money with interest and I shall keep my boat." [14]

Prosperity may have made Fulton pompous. In June 1808, he wrote his friend the artist, Charles Willson Peale: "My steamboat is now in complete operation and works much to my satisfaction, making the voyage from or to New York or Albany, 160 miles, on an average in 35 hours . . . Passengers have been encouraging. Last Saturday she started from New York with 70, which is doing very well for these times when trade has not its usual activity." [15]

A period of prosperity began for the partners. In March 1809, Fulton wrote Benjamin West that *The North River* had paid a clear profit of $16,000 (about $115,000 in today's purchasing power) in 1808. In 1809, she did even better, making upwards of fifty trips, fully booked on most of them.

The steamboat had arrived—so resoundingly that competitors appeared increasingly, both to build rival vessels and to challenge the Livingston-Fulton monopoly in New York State waters. By far the most formidable adversary was Colonel Stevens.

The holders of the monopoly had several times tried to bring Stevens into partnership with them. Fulton in a letter of April 2, 1813 to his friend and first biographer, Cadwallader Colden, stated: "In the winter of 1806–07, my partner, Mr. Livingston, proposed to Mr. Stevens to be a partner with us, to the amount of one-third of every advantage which might accrue from our United States patent and state grants, on condition he would pay one-third of five thousand dollars, the expenses of our experiments, just paying 1666 dollars. He declined this offer." [16]

Fulton and Livingston had again tried to interest Stevens in the spring of 1807 when they needed an additional $1,000. Again

Stevens had refused. If the colonel had accepted either offer, putting his pride in his pocket and taking his wallet out of it, he would have saved himself much worry and much money. Yet, he reasoned that he had already spent more in American experiments than Fulton and Livingston had in French and American ones. He wanted to try out his own machinery, which he considered superior in every way to Fulton's.

One can imagine him pacing on the front lawn of his Hoboken estate in August 1807, confidently expecting the pitiful *North River* to break down or capsize, if not before his eyes at least soon after it had chugged beyond his view. To Stevens' astonishment and chagrin, it ran successfully. During the winter of 1807–1808, he reassessed the situation and offered to join his brother-in-law and Fulton as a partner. They replied to him on January 13, 1808:

> We have consulted together and fully considered your application to become interested with us in our patent rights. We are sorry to find you adress yourself . . . to our fear of not being able to support our pretentions ag't any new patent. Upon this subject be assured we do not feel the slightest apprehension . . . If we did, it would be madness in us to proceed . . .
>
> A long boat, and wheels was, with us . . . the result of expensive and numberous experiments . . . Neither of these entered into your plans . . . nor did you, after we had told you of our intentions to use them, manifest any faith . . . You constantly refused to come into partnership with us . . . because you believed our propelling apparatus was defective . . . Is it right to threaten to fight us with our own weapons?
>
> Considering how large a field the U.S. opens, we had hoped you would have left us the quiet possession of that we had pre-occupied . . . This we flatter ourselves you will be inclined to do, when we first convince you that we have an undisputed right to navigate with steamboats all waters belonging to this state, and conclude, out of mere friendship, by abandoning to you such a portion of our rights as will be more advantageous to you than the partnership you propose.[17]

Although both partners signed the document, the syntax, misspellings and tone clearly indicate that Fulton wrote it. Then followed an angry exchange of letters, one signed by Livingston and

Fulton of special interest because it puts their monopoly position succinctly:

> This state has vested in us exclusively the right to construct, use, etc. all vessels moved by steam—in all the creeks, rivers, bays . . . whatsoever, belonging thereto . . . It enacts further that, if any person . . . shall infringe our right, such person shall, for every such offense, pay to us one hundred pounds and forfeit the boat & engine with her furniture; and gives any court within the state jurisdiction. You will admit that this right is full and clear & unless it is done away by something in the constitution of the U.S., that it is incontrovertible . . . The U.S. have no rights but those derived from the States.[18]

Stevens admitted no such thing. He began construction of a steamboat to be called the *Phoenix* in Hoboken in January 1808, five months after Fulton had first steamed up the Hudson. Significantly, he returned to a low-pressure engine similar to the Watt design in the *North River,* but he built it himself, having failed to buy one from Boulton & Watt. It was still difficult to lure Watt engines out of England. Only three had been allowed to leave the country by this date—one to France for a pumping station at Chantilly, a second to America for Aaron Burr's Manhattan Water Company, and a third to Fulton.

Stevens also swung his paddle wheels over the sides as had his rival. He did not adopt the narrow design of the first *North River,* but Fulton had widened her also in his rebuilding. Stevens was proud of several changes in design detail—special braces for the engine to reduce strain on the hull, an altered air pump, and modified valves.

Before the vessel was one-third finished, Stevens tried again to maneuver himself into the Livingston-Fulton partnership. This time, Livingston offered him a one-fifth partnership, provided that if he used low-pressure steam, paddle wheels, or proportions similar to Fulton's he would acknowledge indebtedness. This really hit the colonel where he lived. He refused.

In April 1808, Stevens launched the *Phoenix,* but did not challenge the monopoly directly in New York State waters. He ran her the thirty miles between Perth Amboy and Paulus Hook, claiming she exceeded five miles per hour, somewhat faster than even the new *North River* could make, on average.

Stevens' wife, who was Livingston's sister, attempted to mediate the quarrel, but both her husband and brother told her to mind her own business. The dispute contained a strange mixture of family affection and professional animosity. For example, when Stevens decided not to challenge the monopoly in New York State waters but to run on the Raritan River between Perth Amboy and New Brunswick, New Jersey, and between Philadelphia and Trenton on the Delaware, Livingston and Fulton were among the subscribers. Yet, when this deal came to nothing, a new arrangement excluded the New Yorkers. Then Fulton and John R. Livingston, Robert's brother, offered to buy a half interest in the *Phoenix*. No deal. In retaliation, Fulton and John Livingston built the *Raritan* as partners and challenged Stevens on the Perth Amboy-New Brunswick run, charging lower rates.

As a result of such competition, Stevens decided to take the *Phoenix* south to the Delaware River as he had originally planned. This was a daring decision because he had to run the paddle steamer in the open ocean. The waves and breezes that were routine for a sailboat in the seas threatened the purchase of paddle wheels upon the water. Commanded by Stevens' favorite son, the twenty-one-year-old Robert Livingston Stevens, the *Phoenix* made the trip in June 1809, the first steamboat voyage on the ocean. It took thirteen days to go from Hoboken to Philadelphia because the *Phoenix* had to anchor in sheltered coves except during the calmest weather. Even so, the waves damaged the paddle wheels twice.

After a successful season in 1809 on the Delaware, Fulton-Livingston and Stevens decided to halt their rivalry. Fulton and Livingston were to have New York, including Lake Champlain; the Brunswick run; and the Ohio and Mississippi, with the general reservation that the Stevens improvements should not be used by them in establishing ferry boats between New York City and the Jersey shore. The colonel was to use anything of Fulton's on the Delaware, the Chesapeake, the Santee, Savannah, and Connecticut rivers, as well as the Providence run. All these were to be Stevens' for the next seven years, provided that, if he had not in that time established steamboats on any particular water, that one should revert to Livingston and Fulton. Also, it was provided that each party should pay the other "a reasonable consideration" for the use of any new improvement.

At first, Fulton had opposed any compromise with Stevens. "I will never admit that he shall navigate the waters of the State of New York," [19] he told his partner once. But he came to see the futility of the rivalry and actively supported agreement by 1809. Then, ironically, Livingston almost ruined everything by proposing an amendment. He demanded Stevens' agreement that he and his heirs and representatives "fully and absolutely recognize" the right of Fulton and Livingston to all inventions and improvements specified in Stevens' patents. Stevens refused to sign this amendment, but the main agreement was implemented.

Although the year 1808 started so well, Livingston would soon write to his brother-in-law (exaggerating just a little perhaps, as he was successfully discouraging a Stevens plan for a boat on Long Island Sound):

> The fact is that I have never received one farthing from steamboats. [With] what I have laid out upon them . . . I begin to tire of all these expenditures & indeed I have not the means of working with hazardous plans of a boat on the Sound, unless I can get something for our state and patent rights . . . which may doubtless be done without making an advance, as the new state-law affords us ample security against intrusions . . .
>
> Something must be done, because the public is complaining that we will neither exert ourselves nor let others do it, without ourselves or our connections having all the boats.[20]

It took Fulton much longer to see the ominous clouds forming overhead.

13 | FULTON, THE MAN OF AFFAIRS

As Fulton's business activities increased from 1808 on, he spent money as fast as or faster than he made it. Like many people who gain wealth late in life, he saw no sense in saving for a rainy day—his rainy days were behind him. He never hesitated to put large sums into his business schemes, into his handsome house and its furnishings in Manhattan, and into the hands of his artist friends and relatives. His brother and sisters had not prospered. Saying that his relations had "in general been unfortunate," he showered loans, barns and cattle upon them.

His largesse grew so copious that it alarmed Harriet. When she protested at his transferring some company shares to his partner, her third cousin, Livingston, she did not understand that her husband had fulfilled a contractual obligation, not followed a caprice. She wrote Livingston that Fulton had sinned against a "defenseless woman." She said that the earnings from the shares would be of little concern to a man as rich as Livingston, but provided her with pocket money four times a year, intimating that she needed it badly.

She understood scarcely the first thing about her husband's multifarious projects. Indeed, they grew more complex by the month. First, he had to manage his shipbuilding facilities. He began to construct a second vessel, *Car of Neptune,* almost a duplicate of the *North River,* during the summer of 1809, but she was not ready until that autumn and did not operate for a full season until 1810. She cost about $25,000, half again as much as the entrepreneur had estimated. With these two boats, he began a New York-Albany service twice a week. At about the same time, he had built the *Raritan* for service on the river of the same name in New Jersey between New York and New Brunswick. Then, he still had to attend to the details of running the *North River*. Thus, he became the first inventor to have more than one ship under steam at the same time.

He had built all the machinery to his own design for the two new vessels in his shops. Usually adept at managing people when his nerves and increasingly precarious health permitted, he played the part of patron, not companion, in dealing with employees. Because this employer-employee relationship was customary in those days, his workers did not resent it. Indeed, they admired him.

Perhaps Fulton, something of an actor in his reactions to an audience, thought this the proper role to play. With peers, he could be effervescent, loquacious, and charming. With people he was trying to impress, he displayed those last qualities plus a proper deference. As the years passed, however, the deference appeared less frequently.

Problems of health bothered Fulton increasingly. His nerves caused stomach disorders, similar to those that had plagued him during his submarine adventures in England. His old lung troubles reappeared by 1812. He may have become tubercular.

His uncertain health did not slow him down; on the contrary, he increased his activities. These included siring four children in rapid succession—Robert Barlow born in 1809, Julia in 1810, Mary Livingston in 1811, and Cornelia Livingston in 1812. His only son, always called Barlow, died childless in 1841, so no direct descendent bears the Fulton name.

By 1810, work had so crowded upon Fulton that he had to turn to Boulton & Watt again for a steam engine:

> In 1804 you constructed for me a steam engine with a 24 Inch Cylinder and a four foot stroke, which engine has for four years past been driving a boat 166 feet long, 18 feet wide drawing 2½ feet of water at the speed of 5 miles an hour on the Hudson River; that is taking the tide for and against the boat her average velocity is 5 miles an hour; This application of your Invention to drive boats, has been, and will ever continue to be of great public utility in this state by carrying passengers Between the Cities of New York and Albany distance 160 miles, the profits have also been such as to induce me to form similar establishments on some of our other riv-

ers. I will therefore esteem it a favour if you will have the goodness
to make for me another engine as soon as possible, the Cylinder to
be 26 inches, the stroke as before 4 feet . . .[1]

Fulton planned a boat to ply the Ohio and Mississippi river system. He and Livingston had applied in 1810 for a monopoly similar to the one they enjoyed in New York. The territory of Louisiana granted them that exclusive privilege on April 19, 1811. As minister to France, Livingston had negotiated the Louisiana purchase. He wielded considerable influence still in that area, enough to close the mouth of the Mississippi to all steamboaters except himself and Fulton. The latter, who ran the day-to-day affairs of their enterprise, delegated the superintendence of this western venture to Nicholas J. Roosevelt.

After several failures in his own metalworking and steamboat ventures, Roosevelt had gone to work for others. He got a job with Benjamin H. Latrobe, an architect and engineer, then building the Philadelphia waterworks. He married Latrobe's daughter, Lydia, in November 1808. Presumably, the father had overcome his objections to steamboaters by then. Earlier, he had called those tinkering with the vessels "crackbrained."

Although Roosevelt had suffered indignities and business reverses in a former association with Livingston, he swallowed his resentment and accepted a contract from the chancellor and Fulton to build a steamboat in Pittsburgh, to be called the *New Orleans*. The paddle wheel, in a relatively new departure, was to be placed at the stern, not over the sides as the other Fulton vessels had them.

The vessel had masts for sails fore and aft, which, when the wind was favorable, were expected to increase her speed. She was one hundred and sixteen feet long and had a beam of twenty feet. Her machinery sat below decks, necessitating a draft of five feet which would prove troublesome time and again on the shallow stretches of the Ohio and Mississippi.

The job began in 1811, on the north bank of the Monongahela River, where most of Pittsburgh then was situated. Timber was cut within a short distance of where the keel was laid, near Boyd's hill, site of Beelen's iron foundry. Boat builders, accustomed to constructing the barges of that day, were available in Pittsburgh, but fifty mechanics had to come from New York.

Troubles dogged the *Orleans,* as she was usually called, even before Roosevelt could complete her. The planned stern wheel design didn't work out, and he had to revert to the more customary side wheels. The Boulton & Watt engine didn't arrive from England, so Fulton had to substitute one he had planned for his third Hudson River boat, the *Paragon.* Other materials remained short, and the hard-pressed Roosevelt never had enough skilled workers. Repeated floods added to the difficulties, pushing the total cost of the vessel to a record $38,000.

Livingston fumed at the expense, and even Fulton decided he had better take the arduous trip by stage from New York to view the situation first hand. Construction of both the *Paragon* and the *Orleans* at the same time had stretched the partners dangerously thin financially. Competition that had mounted on the Hudson and the rising legal costs of defending their monopoly raised the possibility (which would become a certainty) of a financial loss for 1811.

Perhaps Fulton wanted to get away for a while from his eastern troubles. Certainly, he also wanted to see his sisters in Washington, Pennsylvania, twenty-seven miles from Pittsburgh. He had not visited them since 1786. (Nor had he seen his brother, Abraham, who had moved on to Louisville.) What a contrast between Robert and his Fulton relatives! In the quarter century, he had turned himself into an elegant gentleman of the world. His sisters remained country people. The most sophisticated, Polly Fulton Morris, operated a tavern in Washington with her husband, David Morris, the nephew of Benjamin West.

Fulton bloomed in such situations. He was never patronizing, but genuinely entertaining with his stories of Paris, London and New York. His uncritical sisters adored him, as they had all their lives. They restored his spirits. And the *Orleans* at last neared completion. Fulton stayed for the launching, then left for New York, confidant that he, with Roosevelt's help, would score another triumph in opening the Ohio and Mississippi to steamboats.

Roosevelt and his wife, who was pregnant, resolved to take the maiden voyage which began in late September. No other passengers would agree to go. The Pittsburgh crowd on the shore was amazed as the vessel groaned and shuddered her way down the Monongahela. Too excited to sleep, the Roosevelts passed most of the first night on deck, watching the shore slide past, covered then with an almost unbroken forest. On the second day after leaving Pittsburgh, the *Orleans* reached Cincinnati and had to anchor in the river. Levees and wharfs were unknown there in 1811. The townspeople cheered, but also jeered. "You may have visited us in a steamboat—going down river. Coming up is absurd." [2] The *Orleans* reached Louisville October 1, the fourth day after leaving Pittsburgh. The townspeople voiced the same skepticism, especially because the water on the falls of the Ohio at this point was too shallow to permit the vessel to negotiate them safely. The delay proved fortunate in one way. During it, Mrs. Roosevelt became a mother.

Not until the last week in November did the river rise sufficiently to permit the *Orleans* to continue, with Lydia Roosevelt and the baby aboard. Slowly at first, then faster and faster, the ship moved along. She made it! Despite the delay and the birth, the voyage had been pleasant thus far. Then began a time of horror: An earthquake in 1811 had changed the channel in the Mississippi. The pilot became confused. A stove set the ship on fire; fortunately the fire was extinguished before it did irreparable damage. The vessel also had to tie up frequently as close to shore as possible while the sailors turned themselves into lumberjacks to get wood for the boiler. Indians sometimes helped with the fuel-gathering chore and named the steamboat *Penelire* or "fire canoe."

As the *Orleans* descended the river, she passed out of the earthquake region and the going became easier, but she still had to avoid shoals and snags. When she reached Natchez, thousands had assembled to see the sight. There, Roosevelt summoned a clergyman because Mrs. Roosevelt's maid and the ship's captain had fallen in love. They married in Natchez.

The vessel reached New Orleans on January 12, 1812, having traveled 2,000 miles in three and a half months.

The skeptics were partly right. The *Orleans* had insufficient power to move far upstream on the Mississippi. Fulton kept her on a run between the Gulf and Natchez. His second Mississippi boat, the *Vesuvius*, was advertised in 1813 to run between New Orleans and Louisville, but she never made it either. She had to be confined to the New Orleans-Natchez run. The *Orleans* soon struck a snag and sank. The *Vesuvius* languished for nearly one season on a sandbank. Two other western boats that Fulton built, the *Etna* and the *Buffalo,* had scarcely better luck.

The fact was, Fulton's steamboats were designed for rivers flowing toward the Atlantic, not the Mississippi system. His vessels never did ascend from New Orleans to the mouth of the Ohio, let alone Pittsburgh. It remained for other men to accomplish the ascent with more powerful engines to counter the swift current and with special designs to cope with the snags and sand bars. Although Fulton showed the way, here he succeeded neither technically nor financially. In 1814 he wrote that he had grown "tired of distant operations" and that his Ohio company was "alarmed and disgusted with the expenses and the state of their affairs." [3]

Yet even these time and money-devouring pursuits failed to dampen seriously Fulton's sanguine temperament. On June 18, 1811, he wrote Barlow blithely, "My whole time is now occupied in building North River and steam ferry boats, and in an interesting lawsuit to crush twenty-two pirates who have clubbed their purses and copied my boats and have actually started my own invention in opposition to me by running one trip to Albany; her machinery, however, gave way in the first voyage, and she is repairing, which will, I presume, detain her until we obtain an injunction to stop her." [4]

Fulton refers to the *Hope*, launched on May 19, 1811, built for Captain Elihu S. Bunker who maintained a line of sailing sloops between Hudson City and New York. He had recognized that steam would put him out of business eventually, so resolved to adopt the new technology. Winning financial support from some Albany merchants and pirating Fulton-trained workmen, he built two vessels, the *Hope* and the *Perseverance,* launched soon afterward. He proposed to defy the monopoly. His description of the accident that incapacitated the *Hope* on a trip merits repeating because this kind of mishap occurred often in the early days of steamboating:

> While on the passage off Esopus meadows, something appeared to be wrong in the fire-room (which was in charge of a miserable drunken fireman) and the engine moving very slowly. I found on examination that there was not a drop of water in either of the boilers, and that both of them were red-hot, as well as the flues, and must have been so for at least half an hour. The heat was great enough to melt down five solder-joints of steam-pipe, which was made of copper. I immediately started the forcing pump myself, not thinking that there could be any danger in the operation; the effect

of which was a crackling in the boiler as the water met the hot iron, the sound of which was like that often heard in a blacksmith's shop when water is thrown upon a piece of hot iron.[5]

After the *Hope* was repaired, Bunker challenged *The North River* to a race from Albany to New York. On July 27, they began the first steamboat race in American history. The *Hope* made a better start and hogged the center channel, frustrating every attempt by *The North River* to pass. Crowds lined the shores, most people cheering Bunker as the underdog. Fulton had become the symbol of "the pernicious monopoly" in some quarters. At five miles per hour, they sailed so close, stern to bow, that they appeared as one ship with one column of smoke. Two miles above Hudson, *The North River* made use of its lighter draft in a desperate attempt to pass the *Hope* in shallow water. As they pulled abreast of each other, the vessels collided. Although the damage proved superficial, the captains called off the race. They showed less spunk than later Mississippi captains who would risk their passengers, their vessels and their cargoes in wild contests to publicize their vessels.

Fulton and Livingston may have belatedly realized the advantages in publicity that the race gave them because it won extensive newspaper coverage. But at that time, they wanted tamer and more favorable public notice. They each wrote often to the periodicals of the day with dignified accounts of their ventures. For example, Livingston outlined for the January 1812 issue of the *American Medical and Philosophical Register* a history of steamboats, long on the accomplishments of Messrs. Livingston and Fulton, but short on the efforts of any predecessors.

Sometimes they got good, but unexpected publicity from surprising sources. For example, Paul Svinin, a Russian traveler in America about 1811, wrote this of deck life on the *Paragon:*

> Here you see a happy pair of lovers, near them a politician absorbed in the newspapers; there people play chess; in another place a Federalist is arguing hotly with a Democrat, to the sound of a flute or guitar played by a neighbor; in a corner there is a greedy money-chaser annoyed by the children whose clamor distracts him from his accounts; finally dogs and cats add to the fascination . . . It is not a house, but a whole floating town.[6]

All during the summer of 1811, the *Hope* and *Perseverance* steamed in defiance of the monopoly. Fulton and Livingston won in the courts and took possession of the vessels, breaking them up. Although Bunker was driven off the Hudson, he still determined to become a steamship owner. As several times happened in these legal battles, Fulton reached an accommodation with his opponent. He agreed to go into partnership with Bunker for a ship to ply Long Island Sound. Fulton built her in 1813 and Bunker diplomatically allowed her to be named *Fulton*. During her first season in 1814, the War with England made the Sound unsafe, so the monopolists allowed her to run on the Hudson.

With the addition of the *Fulton*, a formidable fleet of steamers now plied the Hudson. *The North River* had grown obsolete and probably had been broken up, but the partners used the *Raritan* now on the Hudson because retaliatory legislation in New Jersey had forced her off the New Brunswick run. In addition, the *Car of Neptune* and the *Paragon* saw regular service between New York and Albany. In 1812, the *Firefly* had come on the scene. She was a small vessel, only 81 feet in length, and although designed for service on the lower river, was used elsewhere as occasion demanded. Fulton now planned vessels for other eastern rivers and in 1814 launched the *Richmond*, intended for the James River in Virginia. The war made it unsafe to send her south, so she too stayed on the Hudson. (He had better luck with the *Washington*, built in the same year as the *Richmond*, and actually steamed her to the Potomac where she saw service.)

So, by 1814 Fulton had six steamboats going up and down the Hudson, not counting others going across it. These were the ferries, which produced another lucrative business. In 1812 and 1813 he designed the *Jersey* and *York* ferries for service between New York and Jersey City. Based on Patrick Miller's double boats, with the paddle wheel put between the twin hulls, the vessels were fat and double-ended—so effective that even today the design has changed little. To move them easily into their slips, Fulton designed a semicircle of pontoons, similar in theory to the piles used now.

His partner in nearly all of his steamboat ventures, Chancellor Livingston, died at Clermont on February 26, 1813. That year Fulton joined with a new partner, William Cutting, to form the New York and Brooklyn Steamboat Ferry Associates, but the engineer had kept his business in the family: Cutting was Harriet's brother-in-law. Fulton sorely missed Livingston and his advice. Although they had had their monopolistic and legal troubles, they had successfully countered any truly serious threats to their public image. Not so with the Brooklyn ferry.

Fulton and Cutting won a lease that allowed them legally to force all other ferries out of business, regardless of their method of locomotion. Furthermore, they could—and did—charge double the customary tolls. Brooklyn citizens held protest meetings. Captains of competing ferries were, of course, enraged, and among them was tough Cornelius Vanderbilt. His undying enmity would play a significant part in the eventual legal and financial collapse of the Fulton-Livingston enterprises.

Yet for a while the Brooklyn ferry commuters were mollified. When the *Nassau* ferry finally began service in 1814, she provided safer, more reliable and faster service than either the former horseboat or sailing vessel. She was easy to load and unload. On one of the first Sundays of her career, she made forty crossings. *The Long Island Star* editorialized: "This is a refinement, a luxury of pleasure unknown to the old world . . . The captain, lordly as old Neptune, drives his splendid car regardless of wind or tide, and is able to tell with certainty the hour of his return. . . ." [7]

Except for the ferries, the Hudson ships grew larger. The original *North River* weighed 100 tons; its rebuilt form, 182.5; the *Car of Neptune,* 295; the *Paragon,* 331, and the *Chancellor Livingston,* 526.

After studying the first Boulton & Watt engine, Fulton built all but one of the rest of his engines himself. Eventually, he simplified the mechanism, eliminating the flywheels because their function could be carried out by the paddle-wheels themselves. He substituted for the clumsy bell-crank engine the more efficient steeple engine he acknowledged he borrowed from Stevens.

With experience, he abandoned the long, narrow design of the first boats. The ratio of length to breadth in the original *North River* was probably 10 to 1; in the rebuilt *North River,* 8 to 1; the *Paragon,* 6 to 1; the *Richmond,* 5 to 1; the *Chancellor Livingston,* 4.7 to 1. For years he stuck with the flat-bottomed, straight-sided and angular pointed shape of the *North River*. But when he designed the *Fulton* for the Long Island Sound, he returned to conventional curved design for heavier seas. When he discovered that it sailed beautifully on the Hudson, he used that design for all his later boats.

The *Chancellor Livingston* included all his final steamboat ideas.

She burned coal, not wood, his one vessel to do so. Although she moved at only six and a half miles per hour, she had much greater carrying capacity than the earlier boats. Her large vistas and big chambers gave a grand effect. The dining saloon filled the entire stern and could become sleeping quarters at night. Upper and lower berths covered with curtains set the style for subsequent steamers and also railroads.

This last vessel named for Livingston cost more than $125,000, almost three and a half times as much as the *New Orleans,* at whose expense the chancellor had been so aghast. (In 1823 the five Fulton ships still steaming on the Hudson had a value, together with the monopoly, of $660,000. They were paying eight percent on that valuation.)

In 1813, Fulton wrote Benjamin West in an almost grandiloquent vein: "My success is extending great benefits over the United States. In one year I shall have steamboats running from Pittsburgh to New Orleans, distance 2300 miles. Also from the Canada frontier on Lake Champlain to Charleston, South Carolina, distance 1500 miles. Taking in all the rivers and bays on the route, fourteen steamboats, or I should say ships for some are 300 tons, have been built in five years. And I am now building thirteen for various waters." [8]

On the route from Canada to the south, he already had several boats operating. From Albany passengers could go by steam to New York City and on to New Brunswick, where a short stage ride put them in Trenton. There they could board the *Phoenix* of Stevens (who had now joined Fulton). Further south, Fulton's plans had less substance. A coach was supposed to go to the Chesapeake, where a steamboat would connect with Baltimore and with another stage that swayed to the Potomac on which the *Washington* (not yet completed) would steam to Norfolk. Fulton planned stages or other means of land transportation to connect four different steamboat runs that could ply in protected waters on down to Atlanta, Georgia—Norfolk to New Bern, North Carolina; Newport Inlet to Little River Inlet, North Carolina; the North Carolina line to Charleston; and Charleston to Savannah. He planned other boats to run on four rivers—the James, the Appomattox, the Cape Fear, and the Peedee. He also contemplated a vessel to steam between Charleston and Sullivan's Island.

To win backers, Fulton advertised in the cities he would serve.

He offered to license companies on the condition that he get half of all profits over ten percent. It would cost extra if he built the vessels. Almost no one accepted his offer. Only one boat that he built himself actually went into the southern waters—the *Washington* on the Potomac. The *Richmond* was intended for the James River, but it never got there because of the War of 1812. The inventor used it himself on the Hudson.

Fulton also returned briefly to an early interest, canals. In a 1797 letter to George Washington he had predicted that "in about 20 years, the canal would run into Lake Erie." He was uncanny in his prediction because the first shovel full of earth for the Erie Canal was dug on July 4, 1817. But he was wrong about the route. He had envisioned Lake Erie connecting with the Susquehanna River. Barlow had predicted the Erie Canal even before Fulton, in the predecessor to *The Columbiad*, "The Vision of Columbus," published in 1787. He has Columbus foreseeing the nation's internal improvements:

> He saw, as widely spreads the unchannell'd plain
> Where inland realms for ages bloom'd in vain,
> Canals, long winding, ope a waterfy flight . .
> Meet the far lakes, the beauteous towns that lave,
> And Hudson joined to broad Ohio's wave.[9]

In 1814, the government of the state of New York asked Fulton to serve on a commission to study the feasibility of a canal between the Mohawk-Hudson River system and Lake Erie. He endorsed the idea, but quit the commission in a huff when it vetoed his proposal to use inclined planes instead of locks. Political differences may also have contributed to the resignation: Fulton was a Jeffersonian Republican, while the leading commissioner, Gouverneur Morris, was a die-hard Federalist. (Politics, however, did not move Fulton deeply, while differences over engineering principles could on occasion arouse him to passion. He defined Republicans flexibly as "those who labor for the public good.")

In 1814, he published his canal proposals independently (really little more than an updating of his original *Treatise on Canals*) as *Advantages of the Proposed Canal from Lake Erie to the Hudson River*. Morris declared his pamphlet "valuable," but he sent it on to the New York legislature "without examining it minutely."

Fulton would have begun railroad ventures too if Livingston

had not vetoed the idea. In 1811, he proposed building a railway to transport coal to Richmond, Virginia. His partner replied on March 1 of that year: "I have before read of your very ingenious propositions as to railway communication. I fear, however, on mature reflection that they will be liable to serious objections, and ultimately more expensive than a canal." Fulton had suggested using wood for rails, which Livingston said "would not last a week." Livingston also objected that "the carriage of condensing water would be very troublesome," [10] indicating that Fulton had not proposed to use the high-pressure engine. Yet the younger man clearly saw the possibility of railroads. Once when he was jolting by stage across New Jersey, a woman passenger asked him about rail travel. "Madam," he replied, "it will come."

In 1814 after Livingston had died, Fulton tried to raise money to build a railroad to exploit coal mines and transport the fuel. He and associates proposed to the New York State legislature that they charter a bank as the financing vehicle. The legislators turned the idea down on the basis that too many banks had already been formed, with inflationary results.

Fulton also corresponded with Littleton W. Tazewell, a Norfolk, Virginia businessman, about the use of the steam engine for flour and saw mills. Nothing came of this possibility either.

As a practical matter, the entrepreneur-engineer had neither the time nor the incentive to look deeply into peripheral activities. He found himself too busy with steamboats, lawsuits, and his first love, weapons of war.

Court battles, particularly, devoured his time.

FULTON IN COURT

A lawyer named Hopkinson paused for dramatic effect in a Trenton, New Jersey, courtroom in 1814. Slowly he lifted to the light a sheet of paper that Robert Fulton had introduced as an original sketch of steamboat mechanisms which he had drawn for Lord Stanhope in 1793. Hopkinson represented Nicholas Roosevelt, now fighting Fulton in a suit for steamboat patents. Thomas Addis Emmet represented the entrepreneur-inventor, but Fulton spent as much time as his counsel before the judge and jury because he usually made such an effective advocate for his own case.

"I would point out to the court," Hopkinson intoned, "that this paper has an American watermark. Is it not strange that Mr. Fulton would use American-made paper when he was in England—and had been for seven years—where cheaper English-made paper was readily available? And something else defies explanation. This sheet of paper was not even manufactured anywhere until 1796 and certainly not in England where it was allegedly employed for the drawings in question in 1793. I ask the court to draw its own conclusions about the veracity of Mr. Fulton."

Emmet hastily recalled the inventor to the witness stand.

"I meant to call the sketch a copy," said Fulton, "a slip of the tongue."

"Shame! Perjury!" came cries from the spectators as the judge gaveled for order.[1]

As it happened, the gaffe only embarrassed Fulton and proved a minor episode and side issue in his major legal effort to keep his tottering monopoly and patent structure from collapsing. Actually, Fulton had written about and sketched steamboat devices in his letter to Stanhope in 1793, but he couldn't find the original at the time and re-sketched from memory. (The original has since come to light.) Obviously, he showed carelessness at best or dishonesty at worst in leaving the impression during his testimony that the copy was an original.

Nevertheless, Emmet and Fulton (and later Emmet and the Fulton-Livingston heirs after Fulton's death in 1815) kept the shaky legal edifice standing until 1823. Before his death in 1813, the chancellor mourned, "Something must be done because the public is complaining (about the monopoly)." Emmet repeatedly warned both Livingston and Fulton in private that they had a weak case. "After the adoption of the Federal Constitution," he wrote in an opinion of January 19, 1811, for their eyes only, "no state legislature had any authority to grant an exclusive right of making, constructing, or employing any machine or invention." [2]

Despite such advice, Fulton and Livingston decided to rely on their original New York monopoly. As time went on, they got deeper and deeper into the legal morass. Eventually they couldn't get out.

Fulton had gone against his earlier beliefs in advocating monopoly. He had once advised Stanhope that, to make an invention lucrative, you should place it "on the liberal footing of other manufactures. If thrown open on a broad basis to all who choose to employ it, free from restrictions and the spirit of monopoly, it will succeed, and those who have embarked their property in it will reimburse themselves with profit." [3] And he had also grandly adopted as his motto when building the submarine in France: "The Liberty of the Seas will be the Happiness of the Earth."

Why, then, did he later turn one hundred and eighty degrees and embrace monopoly and, furthermore, stubbornly defend it for years? Many factors explain his change—the poor patent system in the early 1800s that made monopolies the safest (but not foolproof) way to protect inventions, the related fact that monopoly was a common (although unpopular) way of doing business at that time, and the influence of the Livingstons to whom monopoly was a natural way of life.

To understand this last, we must understand the patroon system that still existed in New York State at this time. In 1629 when the Dutch were trying to colonize the area, the Dutch West India Company established a "Charter of Freedoms and Exemptions"—

the patroon system that was a land monopoly. The charter permitted grants of great river estates to members of the company who would, within four years after accepting the terms of a contract, establish on the lands proffered them settlements of at least fifty persons. The patroonships might extend sixteen miles along one shore of the river or eight miles along both shores "and as far inland as the situation of the occupants will permit." Thus, a feudal system started that tormented the region for two hundred years. The Livingstons were heirs to this system of monopoly, and this helps explain the chancellor's attitude toward the steamboat monopoly. To him it was only right and proper. He convinced Fulton who saw at first hand how the patroons lived, and wanted to live that way, too—especially after his taste of the good life in Barlow's elegant Parisian establishment. So, Fulton changed his philosophy and defended monopoly. He also expected it to further his projects and to make him rich. Tragically the cost of his defense wasted his resources and those of his immediate descendents.

The monopoly's basis rested on New York State's claim that it held jurisdiction over all the waters of the bay and of the Hudson River up to low water mark on the mainland or Jersey shore. Because it was for the navigation of these waters that New York State had granted a monopoly to Livingston and Fulton and because no steamboat could approach New York and trade with the city without going through this stretch of water, we can see how the two monopolists kept everyone else out of a market that was fast becoming the richest in the United States.

New York State's claim violated reason, common sense and even the common law (which assumes that the boundary is in midchannel or in the deepest part). New Jersey had always repudiated New York's claim. In 1806, New York had agreed to the appointment of a joint commission on the dispute to try and reach agreement. The body had to adjourn without a settlement. Why didn't New Jersey appeal to the Supreme Court? Later, that was where the dispute finally would be decided, but in 1806 and for many years, the states were too independent to submit many such cases to the High Court. (Indeed, Chief Justice John Marshall, who would eventually decide this case, frequently complained that he did not have enough to do in the early years of his tenure which began in 1801.)

The first major threat to the monopoly occurred on January 25,

1811, when New Jersey's legislature passed an act declaring that "the citizens of New Jersey have a full and equal right to navigate and have and use vessels and boats upon all the waters lying between the states of New Jersey and New York, in all cases whatever not prohibited by the Constitution, or any law of the United States." [4]

In addition, the act provided that any person whose boat might be seized under the law of New York should have a right to retaliate upon any steamboat belonging to citizens of that state which might come into New Jersey waters. The provisions of this act were even less defensible than were the claims of New York because they encouraged piracy.

The New York legislature retaliated in April 1811 with a law that authorized Livingston and his associates to seize any steamboat infringing on their monopoly. However, it did provide that any steamboat seized should be held until the settlement of the case.

Alarmed now, Fulton withdrew the *Raritan* from New Jersey service, and the partners cautiously refrained from seizing Captain Bunker's *Hope* and *Perseverance* during the season of 1811 even though the New York law permitted them to do so. Fulton asked Stanhope, Boulton & Watt and many other former associates for affidavits, supporting documents, and even for testimony as to his good character. He wrote Stanhope: "The success of my steam boats . . . has been so great that, like every other useful and profitable invention, attempts are now making to evade my patent rights, and deprive me of my mental property. I am therefore under the necessity of collecting all possible evidence of the originality and priority of my invention." [5]

The *Hope* case was tried. The partners won it and won the right to seize and destroy the *Hope* and *Perseverance*. Yet no sooner had they disposed of this than new opposition arose. Colonel Aaron Ogden, former United States Senator and New Jersey member of the 1806 commission on the boundary dispute, petitioned the New York legislature to rescind the monopoly. He lost the resolution to do this by only one vote.

Ogden had entered a partnership with Daniel Dod, an engine builder, to construct a steamboat called the *Sea Horse* with which they hoped to establish a ferry service between Elizabethtown, New Jersey, and New York. The monopoly, of course, prevented them from doing so.

Despite his failure to rescind the monopoly in New York, Ogden had other avenues for attack. He became governor of New Jersey in October 1812, and got a law through his state's legislature in November 1813 granting to himself and Dod the exclusive right to run steamboats on New Jersey waters, which were defined to include the Hudson River. The Livingston-Fulton party girded their loins again and appealed to the New Jersey legislature to rescind the Ogden monopoly. Although Livingston had died by this time, such was the power of his clan in New Jersey, which included the Stevenses by marriage, that the Jersey grant was repealed on February 4, 1815.

Upon the advice of Emmet, the Livingston-Fulton interests settled the matter with Ogden for the time being by selling to him the rights to run ferries for ten years on the route between Elizabethtown and New York. Fulton often reached such accommodations with his legal opponents—not because he was magnanimous, but because he wished to keep his former enemies quiet. He could not afford to withstand a sustained legal attack by any one individual or group. For proof of this, we'll return later to Ogden and the monopoly; for now, let's look to Fulton's legal adventures with his patents.

As we have seen, the young nation's patent system was created specifically to deal with the steamboat and the rival claims of Fitch, Rumsey, and Stevens. In theory, the Secretary of State, the Secretary of War, and the Attorney General were supposed to examine all claims, determine which had priority, and grant the patent on that basis. Jefferson, Knox, and Randolph ducked the issue in the case of the steamboat and granted patents to all three claimants in 1791. In effect, they simply used a registration system, still practiced in some nations today. By 1793, the work of determining priority for patents had become such a burden to the three cabinet members that Congress amended the law to substitute a registration system, in which applications were filed without examination, and rival claimants had to settle their differences somehow, usually in the courts. This arrangement persisted until 1836, when the United States returned to examination as a basis for granting patents, but set up a separate Patent Office under a Commissioner of Patents. From 1802 until 1836, the State Department registered or recorded patents. An official called the Superintendent of Patents had the responsibility for this activity.

It was Fulton's and his heirs' bad luck to have Dr. William

Thornton as the superintendent from 1802 until 1828. He had been associated with John Fitch on the Delaware steamboat, and himself held a patent for an improvement on the boiler for the steamboat. By all odds, Thornton was the least inhibited of any official ever associated with our patent system. He saw nothing wrong in granting patents to himself—a practice now prohibited by law. In one case, after suggesting an improvement to an inventor, he became a joint patentee. For twenty-six years, he ran the Patent Office in a highly proprietary manner, keeping records or not as he liked, making arbitrary decisions about fees—and frequently complaining to the successive Secretaries of State at the low pay and lack of clerical help. His pay never exceeded $1,500 a year, but he managed to get by because he had a comfortable private income. He and James Madison, who first appointed him as superintendent, were next-door neighbors in Georgetown, then on the outskirts of Washington, and they jointly owned a racehorse.

Besides the Madisons, Thornton counted as his personal friends the Washingtons, Franklin, Jefferson, and the Randolph clan. His vigor and personality were such that, in 1814, he was able to persuade the British not to burn the Patent Office.[6]

Probably because Fulton was wary of Thornton, he didn't rush to patent his steamboat plans. At first he thought he could rely on his monopoly for protection. Stevens had his *Phoenix* pounding the waters in 1808, but not on the Hudson. To be a step safer, however, Fulton did decide to apply for his first steamboat patent, and did so on January 1, 1809. He laid most emphasis on the proportions he had devised. Under the registration system, he had no trouble receiving the patent. He also won a second in 1811 for some details of steamboat design.

As competitors began forcing Fulton into court with greater frequency, he found the New York monopoly weakening as a defense. Thus, he had to rely more and more on his patents. They proved weak, too. After consulting with other Philadelphia lawyers, Horace Binney, a leader of the Philadelphia bar in the early nineteenth century, wrote an opinion for John Stevens in 1819. Stevens was associated with the Fulton-Livingston heirs by this time and wanted to know if the patents would hold up.

> The patent of 1809 to Mr. Fulton appears to be confined to his discoveries of inventions in the construction of Boats to run 1 to 6 miles the hour—which inventions consist in ascertaining the pro-

portions between the resistance of the boat, power of the engine, and extent and velocity of the wheels . . . It is not a machine that the Patentee claims to have invented, but tables for the construction of such a machine as to produce a given effect with certainty. This is not a patentable discovery . . .

The patent for the year 1811 for a number of distinct and unconnected inventions . . . such as wheels, wheel-guards, kelsons, coupling-boxes, mode of combining the parts of the engine with the Propeller (sidewheels) etc., etc., all unconnected and independent, is bad because it is contrary to the plain intention of . . . Congress not to permit distinct machines or improvements to be united in one patent. A machine, however numbrous its parts . . . may be patented; but not two or more machines, or two distinct . . . parts of a machine, unless they are claimed as improvement upon (one) already invented and in use.[7]

Even earlier when Fulton had been using his patents in his legal defense, Thornton showed no hesitancy, despite his position, to charge that Fulton's patents were worthless. He concentrated on Fulton's patent statement that boats could not be built to carry a useful cargo at speeds greater than six miles an hour. He trumpeted that Fitch's boat had steamed at eight miles an hour. (By this time no one recalled that Fitch had crammed his vessel with so much machinery that it had room for little else.)

So annoyed did Fulton grow that he wrote Thornton's nominal superior, the Secretary of State, James Monroe at that time: "The case of Dr. Thornton is very simple. If he is an inventor, a genius who can live by his talents, let him do so, but while he is a clerk in the office of the Secretary of State and paid by the public for his services, he should be forbidden to deal in patents, and thereby torment patentees, involving them in vexatious suits. He should have his choice to quit the office or his pernicious practices."[8]

Nobody stopped Thornton. Even Joel Barlow couldn't help Fulton much because, as the recently appointed ambassador to France, he had more than enough problems to absorb him. Thornton, as a doctor (although he never practiced medicine), a self-taught novelist, architect, poet, technologist, construction superintendent, horseman, bon vivant, litigant in many lawsuits, soldier, lapsed Quaker, and busybody, found delight in adding to his range of activities by feuding with the now-prominent Fulton. He wrote about him:

> Finding that Robert Fulton, whose genius and talents I highly respect, has been by some considered as the inventor of the steamboat, I think it a duty to the memory of John Fitch to set forth . . . the following of this opinion; and to show moreover, that if Mr. Fulton has any claim whatever to originality in his steamboat, it must be exceedingly limited.[9]

Thornton also liked to publicize the fact that Fulton had asked for and received an exclusive right to paddle wheels. Of course, the use of these went back to antiquity. While the entrepreneur-inventor derived some advantage from the right to paddle wheels, that also highlighted the shakiness of his claim. With some justice, Thornton pointed out that each item of machinery used on Fulton's steamboat had been patented individually by someone else before him. Thornton continued as Superintendent of Patents and as the enemy of Fulton and his heirs until his death in 1828.

Fulton had to continue fuming against the patent superintendent in private. "A more infamous and outrageous attack on mental property has not disgraced America," he wrote Barlow. "Thornton has been one of the great causes of it." [10]

Yet, some legal and legislative victories buoyed the steamboat partners, notably in 1811, so they kept up their fights in the courts. In addition to the right to seize the *Hope* and *Perseverance*, they also won the new monopoly granted by the New Orleans Territory.

The inheritors of this monopoly would have as much trouble with it as the heirs to the New York monopoly would encounter. Edward Livingston, the chancellor's younger brother, who had migrated to New Orleans following business troubles and a scandal in New York, represented his family's and Fulton's interests. When Henry Shreve defied the monopoly and steamed his *Washington* into New Orleans in 1817, Edward told him, "You deserve well of your country, young man, but we shall be compelled to beat you in the courts if we can." He is said to have offered the river captain half of the western monopoly if he would "instruct his counsel to arrange the business so that a verdict might be found against him." [11] Shreve angrily refused, so off to court they all went. The District Court of Louisiana declared the monopoly illegal in 1819.

But that defeat didn't seem possible in 1811. In that year the partners persuaded the New York legislature to pass a supplementary act providing fines and other legal remedies against anyone maliciously injuring any steamboat on state waters. This had become necessary because flyboat and scow owners collided with the steamers with suspicious frequency. So Fulton often appeared in the state courts on cases of this sort. Although the law helped, it didn't completely stop the "accidents," and for the rest of his life the entrepreneur had to spend a considerable amount of time on these often petty suits.

On the advice of Emmet, the partners avoided big test cases, especially in the federal courts, because he warned them repeatedly that the state monopoly would not hold up there. It conflicted with the patent and interstate commerce provisions of the Constitution.

When they did go to court, Emmet had to rely more on emotion than on law. Normally, Fulton made a superb witness on the stand, looking elegant, handsome, and impressive, giving his testimony with great earnestness and sincerity. At one trial, Emmet summed up by turning from the jury and reproaching Fulton for being too high-minded. In his reliance on the justice of his cause, the good man had spent all his earnings to help others. Emmet continued:

> I admire and applaud you for your readiness to devote to the service of the public the opulence you derive from its grateful remuneration. Let me remind you, however, that you have other and closer ties. I know the pain I am about to give and I see the tears I make you shed! By that love I speak, by that love which, like the light of Heaven is refracted in rays of different strength upon your wife and children; which when collected and combined forms the sunshine of your soul; by that love I do adjure you, provide in time for those dearest objects of your care.

Emmet professed to fear that unscrupulous men might succeed, by raising the cry of monopoly, to get some legislature to "give your property to the winds and your person to your creditors . . . Yes, my friend!—my heart bleeds when I utter it, but I have fearful forebodings that you may hereafter find public faith a broken staff for your support and receive from public gratitude a broken heart for your reward." [12] The public more readily accepted such rhetoric in those days than now. Why did Fulton put up with prodigious expenditures in time, money and effort on legal affairs from about 1808 on? He could have built one hundred steamboats instead of just twenty-one if he had dropped the whole legal mess and concentrated on manufacturing the vessels, leaving the operation of them to others. We have already seen that snobbery prevented him from doing that. Once he had taken the monopolistic course, he found he had a legal bear by the tail, and he couldn't let go. He also put up with the expense because gentlemen went to court almost routinely in those days. Americans in the late eighteenth and early nineteenth centuries were even more litigious than they are now. Many of Fulton's friends (and enemies) were lawyers or at least had a legal education—Barlow, Chancellor Livingston, John Stevens, and many others. American colleges at this time largely trained men to become lawyers, clergymen, or teachers. Hordes of lawyers beat the bushes for business. As a client, Fulton was a counsel's dream.

Fulton probably also put up with the lawsuits because he enjoyed them. In his letters, he many times refers to an "interesting case" engaging his attention. A born actor, he loved to appear on the witness stand. Emmet shrewdly exploited his prominence and popularity which steadily rose from 1812 until his death.

Sometimes Fulton actually invited controversy, as in this exchange with Stevens during one of the periods, in 1812, when the two quarreled:

> I have just been informed [wrote Fulton] . . . that your foreman, and with your knowledge, has been endeavouring to entice some of my workmen, who have gained experience in our shop, to go to work for you at Hoboken. I hope this is not true. But if so, and one man moves from my shop, even by his own voluntary act, I shall instantly insist on all the rights to which I am entitled in law & justice, and which have been encroached on in a manner that cannot be maintained.

Stevens replied:

> Your letter . . . is couched in terms so very offensive that I should not have deemed it incumbent upon me to have returned an answer, were it not that it is necessary and proper I should be informed explicitly what you mean when you say 'I shall instantly insist on all the rights to which I am entitled in law and justice . . .'

When I tell you that I have no knowledge of the circumstances you mention respecting your workmen, be assured, Sir, that the very indecent threat contained in your letter has not, nor ever will have, any influence on my conduct.

Fulton fired back:

That your foreman, or someone else working under you at Hoboken, did attempt to entice some of our workmen away, is a fact; and, I was informed, with your knowledge; hence my letter was a conditional one that, if your knowledge or concurrence, I should insist on rights you have infringed.

Your letter appears to imply that you have not encroached on my rights. It is time, however, that point should be settled by our counsel . . .

Your letters last winter . . . making yourself inventor of steamboats and I a mere cypher, had something in them really offensive and extremely injurious . . .

You had no claims on me, either as relation or friend, yet I came forward and granted you great privileges which I hoped you would at least see the policy of acknowledging and maintaining. But the arduous desire to be thought an inventor has kept you at constant war with Livingston and Fulton, and particularly with me, my interests, and your own through me (a reference to Fulton-Stevens agreements to use each other's patents in certain areas). This you have evinced in your writings, conversations, and acts. But, Sir, here ends writing on this subject, on my part. When justice fails, the law must make up the deficiency.[13]

This quarrel died down and did not lead to suits. Yet Fulton was almost always ready to do battle. (Perhaps the War of 1812 made him unusually bellicose?)

15 FULTON AT WAR

The inventor and his friend Barlow never missed an opportunity to try to sell the submarine-torpedo system to somebody. As long as the British persisted in molesting American shipping and impressing off American vessels into their navy, numerous opportunities developed. Jittery congressmen especially showed interest. A chance to influence many of them arose early in 1810, probably brokered by Barlow.

Fulton delivered a series of lectures in Washington on the "Mechanism, Practice and Effects of Torpedoes" to a substantial number of senators and representatives. In his concluding address on February 17, he was at his best. In his earlier talks he had bombarded them with words, printed matter, drawings, and models. Although he could marshal impressive skills in making technical matters reasonably clear to nontechnical audiences, he nevertheless had to make do with a dry scientific subject. Now he had completed the dullest part of his presentations and could get to the emotional heart of his message.

Tall, still not too heavy, he would pace slowly up and down before his audience, sometimes with his hands clasped behind his back, sometimes with his right thrust in his waistcoat after the manner of Napoleon.

"As men from habits of caution are always distrustful of new inventions,"—he smiled, to exclude his audience from such criticism—"I shall, before I leave you, endeavor to defend this favorite offspring of my scientific pursuits against those persons who may still assail it, by recalling to your recollection some of the strong examples of ignorance opposed to art."

He pushed his long dark hair back from his forehead before continuing: "When a new combination and effect is proposed, and the inventor silently laboring in his cabinet has cleared the way to a fair prospect of success, he receives the common appellation of a projector; the sneers of compassion or contempt which are bestowed upon him, are in proportion to the magnitude of the object which he has in view, and its range beyond the limits of vulgar understandings."

He smiled again, tacitly assuring the senators and congressmen that they had no limits of understanding. Fulton had the gift of making his listeners feel that he and they were above the crowd, capable of rare sensitivity. As he drew them into his magic circle, nobody smiled derisively when he compared himself to Galileo and pointed out the similarity in reactions to the Italian's discoveries and to his own. His words had flown deliberately, even slowly, but they came faster as he stopped his pacing, faced his audience squarely, and summed up in a carrying baritone:

> You who feel sensible of the tyranny which European navies exercise over us, and the embarrassments under which our government and these states now labor in consequence of such tyranny . . . will I hope use your influence to procure the necessary funds for such experiment . . . On my part I volunteer my services to conduct it, and with confidence I promise you the most satisfactory success. But whatever may be your decision, whether you now support this system and carry it into effect, or abandon it to the chances of time; I will never forsake it but with my breath, and I shall hope to see it become the favorite means of protecting the commerce and liberty of my suffering country. Should I sink under the casualties of life, it will be an orphan of the arts which I recommend to the guardianship of my fellow citizens; let them nourish it with the care which I have watched over it for nine years past, and I predict that their recompence will not only be immense in the economy of blood and treasure, but as lasting as this Continent, which it will make our own.[1]

Fulton so impressed Congress that it voted $5,000 in March, 1810 to the Navy Department to conduct experiments. The Secretary of the Navy promptly appointed Commodores Rogers and Chauncey to superintend the operations. By September, Fulton had his models and plans ready, plus a new device for cutting the cables of vessels at anchor. This was quite crude—merely a curved knife, the shaft of which was in the barrel of a gun. The mechanism was supposed to float by means of a buoy against the cable intended to be severed until the knife caught the cable, when a gun lock discharged the gun—hopefully.

By the following month, Fulton was ready for trials in the Brooklyn Navy Yard. He had no inkling of what Commodore Rogers had in store for him. That sly sea dog had ordered Lawrence, captain of the sloop, *Argus,* to resist attack by chains lashed to the cable and booms supporting netting extending down to the sea floor.

Fulton confidently set out in an eight-oar skiff. He had rigged a harpoon gun to fire his mines. With his finger on the trigger, he was about to set off his deadly carcass when, suddenly, his oarsmen backwatered. They encountered not only the floating nets but a pen of logs. When they saw the heavy iron weights hanging from the rigging waiting to plunge down on the skiff and great scythes on the ends of spars ready to sweep back and forth out of portholes, at the height to decapitate the inventor, they had to admit frustration. Fulton returned to shore and admitted that "nautical men and experienced commanders" had met his earlier challenge to show how ships anchored in a calm could escape "total destruction in a few hours."

Even the cable-cutting device didn't work in the demonstration, although he got it to perform better later. However the government's examining commission reported against the invention.

In his rebuttal, Fulton argued that "an invention which will oblige every hostile vessel that enters our ports to guard herself by such means [netting] cannot but be of great importance in a system of defence." He also claimed that he had discovered a means to render "all such kinds of [protective] operations totally useless." [2] If so, no record has survived of what those means consisted of. Anti-torpedo nettings remain a defensive measure to this day.

Despite the failure, Fulton wrote Stanhope, who had invented a type of vessel supposedly immune from torpedoes, that "the ship is very ingenious, but the torpedoes are now so far improved that any plan I have yet seen cannot defend a ship against vigorous attack with them." [3] This proved pure bluff.

He tried once more, this time returning to Bushnell's conception of moving a torpedo carrier awash, not submerged. He proposed a craft he dubbed the *Mute* which would have its top deck covered with iron heavy enough to repel the enemy cannon shot which, because of its lowness in the water, would have to strike obliquely. The vessel would be moved by an armored paddle wheel actuated by one hundred rowers and would creep up on the enemy in darkness.

Fulton never built the full-sized ship, but he did finish a smaller one big enough to hold twelve men under its armor-clad deck. Richard Burdett, captain of HMS *Maidstone,* reported to his superiors in England on June 25, 1814, that "the wonderful turtle boat which has been so long constructing at New York by the celebrated Mr. Fulton" had started to join the American fleet. A gale, however, had driven her onto the Long Island shore. One sailor had been drowned. Captain Burdett rushed to the attack with two warships. "Upon rounding a point of land," he wrote, "I discovered this newly invented machine lying in a small sandy bay, in a wash of the beach, with a vast concourse of people around it, a considerable part of whom were armed militia, who took their stations behind the banks, to the right and left of the turtle boat, which lay on the beach resembling a great whale." [4]

The British cannonaded the beach and then sent marines ashore who drove off the militia and destroyed Fulton's turtle. The inventor did not return to it, perhaps convinced at last that the time was not ripe for either the submarine or the torpedo.

In 1813 he had patented Columbiads, named after Barlow's epic poem. These were guns, intended to fire below the water's surface in such a way that they would hit enemy ships below the water line. They didn't have much force beyond a few yards. Fulton proposed their use in boarding operations. No records indicate that they were used for that or any other purpose. (He was to mount them later, however, at each bow of an invention that did promise to be practical.)

The world's first steam warship grew out of wartime fears and Fulton's two major engineering preoccupations—the steamboat and naval weapons. By combining steam and a war vessel, he sensed that he had at last hit upon a successful idea for a weapon. He threw himself into its development with an almost frantic zeal.

The citizens of New York remembered well the years when the British occupied their city during the Revolutionary War. They doubted that their existing defenses could hold off the English fleet again if it should try to seize their city. They formed the Coast and Harbor Defense Association, with Fulton as a prime mover. He persuaded the association to contract to build the steam war vessel at an expense estimated at $320,000, if Congress would reimburse them if the boat succeeded. Congress voted to do so in March 1814. The unusual, quasi-private arrangement resulted because the Navy already had an extensive shipbuilding program going on. Navy Department officials, cautious in the manner of bureaucrats everywhere, didn't want to gamble $320,000 (which would pay for a conventional sailing frigate) on a chancey steam vessel.

Fulton, of course, had canvased the Navy people for their support before he had adopted the quasi-private method of financing.[5] A letter of January 26, 1814, had failed to convince even Commodore Oliver Hazard Perry, one of the most progressive of the men in the regular Navy, that the government should finance the project outright. At that time, such problems didn't enter the optimistic Fulton's mind.

Adam and Noah Brown, successors to Charles Brown, laid the keel on June 20, 1814, and she was launched, without machinery, on October 29, seven months after the Congressional vote. This would have been fast construction even in peacetime for a vessel this size. For wartime, with the British blockading the coast of the United States, it was remarkable. Fulton worked from dawn to dusk, checking on the work he had subcontracted out in dozens of shops throughout the New York and New Jersey area. He also had to visit Washington several times in 1814 to report progress. On one trip, he kept a diary, a rarity, because he preferred talking to writing. He filled it with sketches, his expenses and technical comments on a steamboat he saw.

Fulton took time out only for the gala launching in October. The river and bay were filled with steamers and vessels of war. Shore batteries boomed in honor of the occasion. Captain David Porter, U.S. Navy, who would command her, wrote to the Secretary of the Navy:

> I have the pleasure to inform you that the *Fulton the First* was this morning safely launched. No one has yet ventured to suggest any improvement that could be made in the vessel, and to use the words of the projector (Fulton), 'I would not alter her if it were in my power to do so.'

> She promises fair to meet our most sanguine expectations, and I do not despair in being able to navigate in her from one extreme of our coast to the other. Her buoyancy astonishes every one, she now draws only eight feet three inches water, and her draft will only be ten feet with all her guns, machinery, stores, and crew on board. The ease with which she can now be towed with a single steamboat renders it certain that her velocity will be sufficiently great to answer every purpose, and the manner it is intended to secure her machinery from the gunner's shot, leaves no apprehension for its safety. I shall use every exertion to prepare her for immediate service; her guns will soon be mounted, and I am assured by Mr. Fulton, that her machinery will be in operation in about six weeks.[6]

On November 21, the Steam Frigate was moved from the wharf of the Browns in the East River, to Fulton's works on the North River to receive the machinery. Two of his commercial steamboats, *Car of Neptune* and *Fulton,* towed her at speeds of three and a half to four miles per hour.

The entrepreneur-inventor proved too optimistic again. The vessel was not completed even by the time of his death the following February. The blockade and dozens of niggling delays and mixups agitated him for the few remaining months of his life.

Fulton had called the ship *Demologos.* The Navy honored him by christening her *Fulton the First.* Most people referred to her as "The Steam Battery" or "The Steam Frigate," (distinctive enough because she was the only such vessel in the world). Fulton borrowed from his ferryboat construction experiences and built her catamaran style, with a double hull, one hundred and sixty-seven feet long, fifty-six feet wide, and thirteen feet deep, measuring 2,475 tons—enormous for that period and a radical departure in warship construction. Anticipating by years the basic principle of the iron-clad, he built her sides and deck of five-foot lumber, which protected a paddle wheel that was also made more secure in its position between the twin hulls.

The inventor likewise planned novel armament for her. Besides the Columbiads, he contemplated thirty thirty-two-pounders shooting red-hot shot. A huge hose connected to a steam pump would spray enemy decks, driving off sailors and so dousing the weaponry that it couldn't function. To move the monster, Fulton stipulated an engine of one hundred and twenty horsepower. In effect, she would be a floating fort, incapable of moving far in the open sea, but effective for harbor defense.

The English had kept themselves informed about the *Demologos,* although the ship grew in the telling. The *Edinburgh Evening Courant* reported:

> Length on deck, three hundred feet; breadth two hundred feet; thickness of her sides, thirteen feet of alternate oak plank and cork wood—carries forty-four guns, four of which are hundred pounders; quarter-deck and forecastle guns, forty-four pounders; and further to annoy an enemy attempting to board, can discharge one hundred gallons of boiling water in a minute, and by mechanism, brandishes three hundred cutlasses with the utmost regularity over her gunwales; works also an equal number of heavy iron pikes of great length, darting them from her sides with prodigious force, and withdrawing them every quarter of a minute! [7]

Lord Napier probably echoed the opinion of most navy men anywhere when he said in the British Parliament: "When we enter His Majesty's naval service and face the chances of war, we go prepared to be hacked to pieces by cutlasses, to be riddled with bullets, or to be blown to bits by shell and shot; but, Mr. Speaker, we do not go prepared to be boiled alive." [8]

So alarmed were the British that they put spies to trail Fulton. They even staged a commando raid on a house on Long Island where he had intended to spend a night. Fortunately, he was delayed and was not present when the attack came. The abortive attempt drew wide publicity in New York. Fulton was hailed as a savior.

As engineer for the project, he was driven nearly to distraction during the winter of 1814–15, moving from shop to shop supervising the work he had subcontracted out. His death delayed but did not stop construction. (However, it did halt the work on a similar vessel for Baltimore.)

Charles Stoudinger, Fulton's superintendent, completed the *Demologos.* This in itself holds great significance. So well had the engineer trained his employees and so carefully had he planned and recorded every step of his design, that others could carry on. Although such an approach to manufacture is routine today, it was not common in 1815. The entrepreneur usually carried most of the plans around in his head, with no one else privy to the whole design. As a result, any mechanical project almost always died with the projector. True, many of the actual specifications of Fulton vessels don't jibe precisely with those published. This resulted

because he made changes as he went along, but he kept his people
informed. In fact, they often suggested modifications. Always the
pragmatist, Fulton listened, adopted, and adapted. (No wonder
his employees liked him.)

In June 1815, the Fulton-trained workers had put the *Demologos'* engine on board, and she was complete enough to try the
propulsion machinery. On June 1 at ten in the morning, propelled by her own steam machinery, she left the wharf near the
Brooklyn ferry and proceeded majestically into the river. Though
a stiff breeze from the south blew directly ahead, she stemmed the
current with ease. She sailed by the forts and saluted them with
her thirty-two-pound guns. Her speed exceeded the most optimistic expectations, and the firing of the guns did not disturb the
smooth running of the steam engine.

On July 4 of the same year, she made a passage to the ocean
and back and went fifty-three miles in eight hours and twenty
minutes. Another trial came in September, this time with the
weight of her whole armament on board. She achieved an average
of five and a half miles per hour, comfortably above the four
miles per hour that Fulton had guaranteed.

But the Treaty of Ghent had ended the War of 1812 on February 16, 1815, so the Navy and the City of New York did not need
the *Demologos.* She was taken to the Brooklyn Navy Yard and
moored on the flats abreast of that station. The Navy officially
commissioned her on June 18, 1817. With President James Madison aboard, she made a short trip to the Narrows and Staten
Island and was again laid up.

In 1821 when her guns and machinery were removed, it was
discovered that she was rotting. From then on the once mighty
Fulton the First became a receiving ship—until June 4, 1829 when
she blew up, killing twenty-four men, one woman, and injuring
nineteen. Two and a half barrels of condemned gunpowder
caused the explosion, completely destroying her. She cost
$290,398.12, a stupendous sum for those days, but for once Fulton had not run over his budget of $320,000 in building her.
(This may have been because another person superintended completing the vessel.)

The ignominious fate of becoming a receiving ship came despite the recommendation of three civilian commissioners who
represented the Coast and Harbor Defense Association in build-

ing her. Henry Rutgers, Samuel L. Mitchel, and Thomas Morris recommended unsuccessfully on December 28, 1815, to the Secretary of Navy:

> After so much has been done, and with such encouraging results, it becomes the Commissioners to recommend that the steam frigate be officered and manned for discipline and practice. A discreet commander, with a selected crew, could acquire experience in the mode of navigating this peculiar vessel . . . It is highly important that a portion of seamen and marines should be versed in the order and economy of the steam frigate. They will augment, diffuse, and perpetuate knowledge. When, in process of time, another war shall call for more structures of this kind, men, regularly trained to her tactics, may be dispatched to the several stations where they may be wanted.[9]

Another coast-defense steamer replaced *Fulton the First,* also named *Fulton,* but she was not built until 1837–38. Even so, it would be many more years before the steam warship would come into her own. Fulton could solve mechanical problems, but he never fully mastered bureaucratic hurdles.

No one will ever know how effective *Demologos—Fulton the First* would have been in war because she never saw action. However, she played a significant part in contributing to a new image of America in Europe after the War of 1812. The newspapers and military intelligence had widely publicized her in Europe. The fact that most such reports exaggerated her formidability only increased the psychological importance of the vessel.

Europeans of 1814 and 1815 regarded America in many ways, but few thought her capable of designing, let alone building, such a deadly weapon of war. Europe was amazed and England chagrined that a nation not yet 30 years old, whom most considered a technological and military backwater, dared to pit her dozen little warships against Great Britain's 1,000-ship navy. Most Europeans believed, with some justice, that France had won the Revolution for America. But now America was going alone—and going quite well.

The War of 1812 started an uphill, downhill fight that ended not only in clearing the seas for American commerce, but in clearing the western frontier for expansion and the whole nation for a technological explosion. Fulton's warship and commercial vessels seemed the answer to the tremendous outcry for better transportation and a land safe from invasion from any enemies— European or Indian. Now, agriculture, commerce, and industry were adding their own demands for a better way to market. Along the eastern seaboard, many of the new mills and factories that had sprung up to fill wartime needs had shown so much Yankee enterprise that they produced goods enough to replace those formerly imported from Europe, plus sufficient more for a burgeoning export trade as well.

In the South, Eli Whitney's cotton gin had placed calico within the reach of every pocketbook. In Pennsylvania, the first arkful of something called anthracite had been loaded at Mauch Chunk and brought laboriously down from the hills—though not many believed much future existed in trying to burn rock. And on the flatlands west of the Alleghenies, wheat stood a good chance of becoming profitable now that England was willing to pay $10 a barrel for it.

People and commodities were on the move. Steamboats were ready to carry and protect them.

16 | THE GIANTS DEPART

Within two months during the winter of 1812–1813, death took Robert Fulton's closest friend, Joel Barlow, and his most contant business partner, Robert Livingston.

Although the American ambassador to France died in December, the engineer didn't learn the bad news until shortly before Livingston passed away the following February. The double deaths hit Fulton hard. He had been ill intermittently with a recurrence of his old lung trouble during much of 1812 and still had not recovered his full strength. He had always regarded Barlow as virtually a contemporary, although he was eleven years older. Now he had died from lung trouble, from pleuro-pneumonia aggravated by exposure. Fulton must have felt the icy chill of premonition.

Barlow had been trying unsuccessfully for an audience with Napoleon. The Americans wanted a commercial treaty with France and restitution for the American shipping the French had seized during the war with Britain. Upon the advice of the French Foreign Office, Barlow had gone to the emperor's intended winter headquarters at Vilna, Russia, to win Napoleon's approval. He waited fruitlessly.

During the demoralized French retreat from Russia, the fleeing emperor passed him on the road to Warsaw. Accompanied by his secretary-nephew, Thomas Barlow, and by a French official, the fifty-eight-year-old Barlow headed toward Vienna in bitter cold. He never got past a peasant's cottage at Zarnowiec near Cracow in Poland. He died on December 26, 1812. His grave remains there to this day—that of the first American diplomat to die in the line of duty.

Although the chancellor's passing on February 26, 1813 at his home in Clermont at the age of sixty-seven was not unexpected, Fulton sorely missed him too. He relied on him especially for support in the increasingly complex legal and business decisions that the partnership faced. From now on, his judgments made without the chancellor's advice would turn out less well.

Characteristically, upon hearing the Barlow news, Fulton immediately invited Ruth Barlow to stay as long as she wished with his family in New York. Some of the correspondence of that period hints that Harriet Fulton either did not concur in the invitation or was not even consulted.

The widow in Paris almost died herself of grief. The legation clerk advised Fulton that her plight was pitiful. She survived and was able to write President Madison: "You will doubtless be informed before receiving this of the dreadful event which has deprived me of the best of husbands & my country of a zealous & devoted friend. Borne down by this cruel, this unexpected stroke, I know not how to get fortitude to support my sinking health." By summer, nevertheless, she could travel and packed for the long voyage home. She left with her sister and nephew in early September.[1]

Ruth Barlow accepted Fulton's invitation and passed the winter of 1813–14 in New York with his family while she gradually recovered. Fulton was so busy that he probably didn't have much time to spend with his beloved "Ruthinda." He was a sensitive man, however, and the tensions between his wife and friend could not have escaped him.

When Ruth returned to Kalorama, all three must have felt relief. Never in Ruth's correspondence does she refer to Fulton's wife with affection or even friendship. Almost no reference of any sort to Ruth Barlow surfaces in the correspondence of Harriet Fulton that survives.

For his part, Fulton had changed. In the year since the double deaths, he seemed to have doubled his activity. His friends noted unaccustomed outbursts of temper. He had lost some of his genial disposition and had gained weight. He complained of stomach trouble, similar to what he had suffered in England during the submarine adventure. Something was driving him, possibly the knowledge that he had little time left in which to accomplish much.

Like most people in the early nineteenth century, he called for the doctor only as a last resort. Yet he must have felt himself at or near the last resort on several occasions during the last two years of his life. His account books show frequent payments to Dr. John W. Francis. While the doctor may also have cared for Harriet and the children, he was Fulton's personal physician, and attended at the time of his death.

By this time, Fulton had become a formidable figure in the business, social, intellectual and popular world of the early nineteenth century. If there had been a Jet Set then, he would have belonged. Besides his business interests and political connections through Livingston and Barlow, he was the prime mover in New York City's Coast and Harbor Defense Association and through this the darling of the citizens of the city. He was a director of the American Academy of Fine Arts, a fellow of the American Philosophical Society and of the New-York Historical Society and was one of the founders of the Literary and Philosophical Society of New York. A patron of the arts, he was also often honored as America's leading engineer.

While not so universally regarded as an inventor and genius, a substantial number of his fellow Americans accepted him in those capacities, too, because he so charmingly and disingenuously claimed such attributes. He looked and acted the part of leading citizen. Even his increased weight added to his formidable presence.

In January 1815, Fulton's lawyer, Thomas A. Emmet, was trying to win the repeal of a retaliatory New Jersey law that gave a steamboat monopoly to his business enemies. Emmet brought his client along as his chief witness, as he often did. Although the weather was bad and the engineer had not completely recovered from a cold, Fulton had agreed to come because this case was unusually important. Unfortunately, he had to wait for a lengthy period in a badly heated hall. The actor in him let none of the legislators realize he felt ill as he testified in his customary impressive manner.

After the hearings, he and Emmet hurried back to New York City. Ice in the Hudson had delayed the ferry, so he inspected his shops on the Jersey shore for three cold hours. At last the ferry crunched its way to the shore. In a hurry to reach it, Fulton led Emmet on a short cut across the ice. The lawyer fell through. Fulton rescued him. His physician, Francis, wrote, "In this moment of great peril, Mr. Fulton was exceedingly agitated, at the same time that his exertions to save his friend left him much exhausted." [2] When he at last reached home, he could scarcely speak.

Although he went to bed, in insisted on leaving it only three days later because another in the interminable series of hitches with the *Demologos'* construction had developed. Both his doctor and wife pleaded with him to stay indoors. He pleasantly but definitely ignored their entreaties. (For years the only advice he had consistently followed had been Barlow's or Livingston's—now they were gone.) He ordered his carriage and departed for the warship. Fulton returned soon, desperately sick this time. He lingered on, but grew steadily weaker. Finally his lungs could function no longer. He died on February 23, just two years to the month after the news had reached him of the death of his two closest associates.

The day of his funeral, Friday, February 24, dawned cold and gray. The temperature stood at 22 degrees Fahrenheit at 8:00 A.M. Wet snow began falling later as the temperature rose a few degrees.[3] The weather fitted the mood in New York City. Minute guns boomed in the harbor from the *Demologos*. West Battery minute guns also sounded from the time the funeral procession started from his home at Number 1 State Street and advanced north a half dozen blocks to Trinity Church.

Officers of federal, state and local governments and many of the prominent city residents attended the funeral service in his home. Offices and shops closed as a sign of respect.

The body, in a lead coffin, enclosed in mahogany, identifying him with a plate engraved only with his name and age, was destined for the Livingston family vault in Trinity Churchyard. Livingstons dominated the funeral and procession. If Fulton's own brother or two surviving sisters attended, no contemporary accounts record their presence. They probably had neither the money nor the time to afford the trip from their homes on the western frontier.

Fulton had not yet reached his fiftieth birthday.

He thus, like his father, left a young widow with small children. Robert and Harriet's four were only six, five, four, and three years old. Fulton at least thought he was leaving a large estate. By his will dated December 13, 1814, he provided Harriet with $9,000 per year during her lifetime, the income to derive from the profits on his steamboats or from other property. He also left her all his household effects. Each child was to receive $500 per year until each had reached the age of twelve. After that, they were to get $1,000 each up to the age of twenty-one. He made bequests to his brother and sisters, forgiving all loans he had made to them at various times. The remainder of his estate went in trust for his children, each to receive his or her share upon attaining the age of twenty-one. In the case of the death of all his children before that of his wife, half of his estate was to go to the "promotion of an Academy of Fine Arts for historical and scientific paintings." The other half went to his wife outright. He left to Ruth Barlow all the copies of her husband's poem, the *Columbiad*, which had become Fulton's property. Further, he directed that the money owing to him from Barlow's estate would be left to Ruth's option concerning repayment during her lifetime. He named Harriet and her brother-in-law, William Cutting, executors and trustees.

Most of the provisions in the will proved academic. Litigations, mistakes and misconduct of some of his agents hired to build boats greatly reduced his estate, even though the Fulton-Livingston vessels plying the Hudson still yielded a profit of about $50,000 a year as late as 1823. Livingston had left no son, and Fulton's was much too young to carry on. At the time of his death, the United States government owed him $100,000 for his services and expenditures on the *Demologos* and for the use of his vessel, the *Vesuvius,* at the battle of New Orleans. And unexpected debts also surfaced after his death. For example, Fulton owed $8,000 to the estate of the chancellor. Old bills kept cropping up that he allegedly still owed from his days in England and France.

Harriet and the children had to give up the house at 1 State Street in New York City and move in with her parents at Teviotdale, north of Germantown in Columbia County.

Soon afterwards she married an Englishman, Charles Augustus Dale, who was more interested in horses than steamboats. On a wager, he bet that he could drive a pair of horses from New York City to Teviotdale faster than it would take the steamboat to sail from New York to the Livingston dock, about a hundred miles. He won, but the escapade killed one of his horses.

In 1825, Harriet Dale and others petitioned the New York Assembly to form a bank. They proposed to set aside the interest on the sum of $70,000 raised in forming the institution for the use and benefit of the heirs of Robert Fulton who, the petition stated, were "utterly destitute of support." The petition failed.

In 1829, the twenty-year-old son, Robert Barlow Fulton, wrote this pathetic letter to an uncle after his mother had died and he had learned that his stepfather, Charles Dale, was trying to sell his life interest in Teviotdale:

> While Mr. Dale was yet in this country, it was frequently told me that when he relinquished all control over my sisters, the family would come forward to assist them. Since he has returned to his native country, it is true my sisters have found a home with their relations, but you doubtless are well aware that that is not the only requisite. Money they have none, even to buy necessary garments . . . What I can save from my small salary is by no means sufficient for even one of them. I do not wish to remind you of the manner in which they were brought up. They merely ask for necessaries & heaven knows they are in want . . .
>
> For them t'is I ask, not a job for myself. When I consider their present situation, dependent on others' bounty for all the necessaries of life, even setting aside comforts, I almost wish!—I know not what. Oh my dear Sir, may they look to you for a friend? [4]

Fulton's friends next tried to recover the $100,000 that the government owed him at his death. Finally, in April 1836, Congress referred their petition to the Secretary of the Navy. Nine months later, he reported his opinion that the United States did indeed owe the Fulton estate $100,000. But it took another nine years for a bill to pass, in July 1846, to repay the estate—$76,300, not $100,000. By this time, more than thirty-one years after Robert Fulton's death, his son—as well as his wife—had already died. His eldest daughter, Julia, would die two years later. Two other daughters, Mary and Cornelia, would live until 1860 and 1883 respectively. The three daughters had all married, and had borne nine Fulton grandchildren, but none carried the Fulton surname.[5] The only son, Robert Barlow, never married.

What did Robert Fulton's life achieve in the fields of invention, engineering, technology, and transportation?

By 1815, Fulton had designed twenty-one successful steamboats. On the Hudson, he had the seven already mentioned, plus the *Chancellor Livingston* under construction at his death. For the Mississippi, he built four. In addition to the three ferries, he also constructed the *Camden* in 1812 for service to that New Jersey city. Planned but not built at his death was *The Olive Branch*, designed to try again on the New Jersey rivers. He designed the *Connecticut* for Long Island Sound traffic. Besides the *Washington* for the Potomac, he designed a Russian vessel and the *Demologos*, the world's first steam warship.

Fulton tried to spread the steamboat abroad. In 1811 he began negotiations for a monopoly in Russia providing that he would within three years run a vessel between St. Petersburg (Leningrad) and Kronstadt through a sheltered arm of the Gulf of Finland. He began to build the *Empress of Russia* (but she was unfinished at his death which also saw the collapse of the agreement). In 1812, he attempted to involve Thomas Law, a former official of the East India Company, in a steamboat for India. Nothing came of this until eight years later when English interests introduced the vessel on the sub-continent.

As an inventor, Fulton was largely a compiler, adapter, and sometimes improver of others' discoveries. He considered this legitimate invention. When rivals disagreed—and especially if they disagreed in a public forum—he would rebut them by a letter which usually found its way into the columns of some well-regarded periodical. A typical case occurred in 1814 when Aaron Ogden attacked Fulton's originality before a committee of the New York State legislature studying the steamboat monopoly.

"If at Albany," wrote Fulton, "you have impressed the committee with a belief that I could be so base as to pirate the labors of others and present them to my liberal countrymen as my own, you have done an unjust and ungenerous deed, which would make the cheek of rigid honor blush.

"Suppose you were to collect a basket of old ballads and bad verse without ideas, but rhyming and containing the twenty-four letters of the alphabet, could you not from those parts used by Pope prove that he did not conceive or invent the *Dunciad* or *Essays on Man* and *Criticism?*" [6]

He avoided ever actually claiming that he had invented the steamboat. Rather, he said he had found the right combinations to build a successful vessel. Thus, he could and did claim, with justice, that he had invented the first commercially practical

steamboat. He always claimed that he had invented the submarine, but we have seen that more credit belongs to Bushnell than
to him. Until virtually the day of his death, he found himself involved in an embarrassing dispute with Nathaniel Cutting over his
charge to the New Jersey House of Assembly that Fulton had
pirated Samuel Cartwright's invention of a rope-making machine,
and then had sold it to Nathaniel (no relation to Harriet Fulton's
brother-in-law William Cutting) as his own.

Fulton was really more an engineer than an inventor, at a time
when only a handful of people practiced civil engineering in
America and relatively few in England, the true birthplace of the
profession. Yet even in this field Fulton had his detractors. John
Rennie, one of England's first great engineers, said: "I consider
Fulton, with whom I was personally acquainted, a man of very
slender abilities though possessing much self confidence and consummate impudence." [7]

Fulton, however, did advance engineering in America. He
brought discipline and order to the art of making mechanical
devices work better. Each of his steamboats showed improvements
over its predecessor because he kept records, made models, and
steadily tried to simplify his apparatus.

Above all, Fulton was one of the first technologists in the world,
a word not even coined during his time, but now defined as one
who specializes in gathering the technical means of achieving a
practical purpose. Many commentators have criticized Fulton's
practice of picking here and borrowing there to come up with the
steamboat. Yet this gathering process was urgently needed in the
case of that vehicle. At least twenty inspired inventors had tin-
kered with it before Fulton and had failed to come up with anything practical. He did. Furthermore, he had sold his idea to the
public, an achievement in which Fitch especially had failed.

Rennie sneered at Fulton's "impudence," and many others criticized his promotional activities. To achieve a practical result, however, requires selling the idea to users. This Fulton did many
times—with the steamboat, the submarine, even the rope-making
machine. He could sell heads of state, legislators, hard-headed
businessmen, and just common people, if only to pay to ride on
his steam vessels.

His selling methods ranged widely—from personal persuasion,
to lectures, to advertisements in newspapers, to pamphlets, to ar-

ticles placed in periodicals in much the manner that a modern public relations man operates.

Sometimes his selling methods backfired, notably in his efforts to guarantee sales through his monopolies. Yet the charge that steamboat transportation would have started sooner and spread faster if his monopoly had not existed misses the mark. Monopolies were a virtual necessity in his time because the patent system was so porous. Fitch, Rumsey, and Stevens all had monopolies for various areas at various times. None of them succeeded in commercializing the steamboat, either with or without a monopoly, before he did. Nor did any successfully operate steamboats in non-monopoly areas before he did.

The factor that limited the spread of the steamboat at least as much as the monopoly from 1807 to about 1820 was the lack of knowledge about how to build the vessel—in other words, a shortage of technology. Stevens, Bunker, and later rivals got their steamboats constructed by pirating Fulton's employees and even, in some cases, designs. This is why he threatened them with court action, and actually sued once or twice, despite the fact that they didn't operate the boats in his monopoly-protected waters.

Fulton had trained his employees carefully, and they had all learned refinements in the art of steamboat construction from each successive vessel. His costs rose with each new boat partly because he had to keep offering his workers higher pay to hold them. Fulton's snobbishness kept him from advertising the fact, but he was far and away the most accomplished and experienced steamboat builder in the world by 1815.

A good case can be made that Fulton didn't even need a monopoly by 1815 to have kept the leadership for years as both a steamboat builder and operator, at least on the Hudson. As events proved, he and his successors would have been wiser to have abandoned the monopoly fight no later than 1820 and concentrated on improving their technology. That would have offered far more protection.

Instead, Fulton's successors let their technological advantage erode away and concentrated on the court fight more than on anything else. In a strange epiloque, Fulton became an unwitting father of free American interstate commerce because his successors lost a landmark case that established it.

EPILOGUE

The case of *Gibbons vs. Ogden,* one of the most momentous decisions ever handed down by the United States Supreme Court, arose directly out of the Fulton-Livingston steamboat monopoly in New York State. Aaron Ogden had purchased rights to operate in New York waters from the Fulton-Livingston monopoly. Thomas Gibbons challenged those rights.

Ogden was an enormous man, with a truculent temperament. Gibbons possessed a disposition even more contentious than Ogden's. He had been born in Georgia and had remained a Loyalist during the Revolution. He survived that stigma because his brother and father had proved themselves ardent patriots. Nevertheless, he evidently felt it wise to move to New Jersey in 1811. He had thrived there as a lawyer, although an enemy described him this way: "His soul is faction and his life has been a scene of political corruption." [1]

To help thwart Ogden and his steamboat, the *Atlanta,* Gibbons had hired as captain of his steamboat, *Bellona,* none other than Cornelius Vanderbilt whom Fulton had driven off the Brooklyn sail ferry run. Vanderbilt hated the monopoly (as long as it wasn't his own) and became Gibbons' zealous ally in concocting ways to outwit it. Colonel John Stevens became another convert to Gibbons' cause.

While Gibbons fought the issue through the courts, Vanderbilt fought on the front lines: he cut fares. He built a secret compartment in the *Bellona* where he hid when New York State officials came aboard to try to arrest him. Whenever they did catch him, he could produce papers that indicated a third party had leased the ship from Gibbons.

Ironically, the two parties to the suit came late on the scene, had at one time even been partners, and never had close business or personal relationships with the progenitors of the entire tangle, Robert Fulton and Robert R. Livingston.

In February 1824, the Supreme Court finally heard the case. Daniel Webster and Attorney-General William Wirt represented Gibbons. Thomas Emmet and T. J. Oakley represented Ogden. Chief Justice Marshall presided on the bench to decide whether Gibbons, Stevens or any other man might run steamboats on the Hudson—or whether the Fulton-Livingston successors could keep the river as a private waterway. Marshall's subsequent fame as a jurist stems importantly from his decision in this case. Daniel Webster won his first national fame as the victorious lawyer.

In his decision, Marshall contradicted the Ogden contention that the powers of the central government, as representing a grant by sovereignties or states, must be strictly construed. He said: "The Constitution contains an enumeration of powers expressly granted by the people to their government, and there is not a word in it which lends any countenance to the idea that these powers should be strictly interpreted." The men who framed the Constitution "must be understood to have used words in their natural sense and to have intended what they said." If, from the imperfections of language, any doubt arose, then "the known purpose of the instrument should control the construction put upon the phraseology." As Marshall saw it, the grant made by the several states to the federal government does not "convey power which might be beneficial to the grantor if retained by himself . . . but is an investment of power, for the general advantage, in the hands of agents selected for the purpose; which power can never be exercised by the people themselves but must be placed in the hands of agents or remain dormant."

With these principles explained, Marshall outlined their application to the Gibbons vs. Ogden case: "Commerce undoubtedly is traffic. But it is something more—it is intercourse. The power of Congress is complete in itself, exercisable to its utmost extent, without limitations other than are prescribed by the Constitution itself. If the sovereignty of Congress is plenary as to these objects, the power over commerce with other nations and among the several states is vested in Congress as absolutely as in a single government." Such power could not be confined by any mere question of state lines but must act everywhere. "It may, of consequence, pass the jurisdictional line of New York and act upon the very waters to which the prohibition now under consideration applies."

Any state enactment, Marshall argued, must give way to the supremacy conferred by the Constitution upon acts of Congress. He demolished the Ogden counsel's argument that Gibbons merely had the right to operate coastwise, by drawing no distinction between "up and down" and "across." (In so doing, he included the Hudson in the Atlantic Ocean.) [2]

Marshall had cut the knot. His decision and the years of litigation left Ogden bankrupt. Although Gibbons ended a millionaire, the long contentions had ruined his health to such an extent that he merged his interests with another steamboat operator, into the Union Line. And of course the decision left the heirs of Robert Fulton in financial ruins.

Yet in a larger sense, the Marshall verdict unleashed Fulton's commercial steamboat. It made possible the golden age of the vessel, especially on the Mississippi. The decision's ultimate effects range even more broadly. It made it possible for the railroad, the telegraph, the telephone, the radio and television signals, the oil and the gas pipelines, and the airplane to move freely across state borders—all because of protection given them in the decision of *Gibbons vs. Ogden.*

NOTES

CHAPTER 1

1. Until farmers generally began to use lime in the 1830's, this area around Lancaster, Pennsylvania, remained unfertile.

2. Fulton was a place name in both northern England and southern Scotland. During the thirteenth century when population increased markedly in both countries, adult males adopted a second designation for differentiation purposes. Often they took the name of the place where they lived or owned land. Hence, Fulton became a common surname. Speculation about Robert Fulton's ancestry is made nearly meaningless by the fact that many unrelated or distantly related people bore the same last name. All that can be known is that thousands of farmers or would-be farmers from Scotland and England emigrated to Ireland at the end of the sixteenth and the beginning of the seventeenth centuries because the English started a new policy of dealing with subjugated territories in Ireland. The government offered land grants to English and Scots settlers to displace the native Irish. Several families named Fulton emigrated to Ulster. By the end of the seventeenth century, this area had become the most prosperous in Ireland, so successful in farming and manufactures that the alarmed English levied ruinous tariffs against its crops and products. This led to extensive migration of the Scotch-Irish to America early in the eighteenth century.

3. The account books of a Philadelphia merchant indicate that in less than two years Robert Fulton, Sr., purchased goods worth £375 and once purchased five hogsheads of rum in two months.

4. Smith family lore had it that their name originally had been McDonnell. Just before the Battle of Boyne in Ireland, King William III's horse lost a shoe in the McDonnell neighborhood. As a farmer, the head of the house was a jack-of-all-trades, including blacksmithing. He reshod the horse and was ever after called McDonnell the Smith. His descendants kept Smith, but dropped McDonnell.

5. The Fulton National Bank now stands on this site.

6. This land originally belonged to a James Gillespie.

7. Flexner, James T., *Steamboats Come True, American Inventors in Action*

(New York: The Viking Press, 1944), p. 111. Although Flexner focuses on the steamboat and most of the people connected with its invention, he includes so much information on Fulton that the book rates as a biography of him.

8. Sutcliffe, Alice Crary, *Robert Fulton and the* Clermont (New York: Century, 1909), p. 21. A biography written by the great-granddaughter of Fulton, this book concentrates on his character and personality. It contains much information and many letters not previously published.

9. Ibid., p. 14.

10. Ibid., p. 17.

11. Dickinson, Henry W., *Robert Fulton, Engineer and Artist, His Life and Works* (London: John Lane, 1913), p. 14. Letter dated April 14, 1789 in possession of Mrs. Frank Semple of Sewickley, Pa. In this and other quotations from Fulton's letters, spelling and punctuation are retained as they originally appeared.

Dickinson was a distinguished British historian of engineering. Not surprisingly, he concentrated on Fulton's engineering accomplishments and is excellent in that area.

12. No evidence holds up that the West and Fulton families knew each other in Lancaster. While Benjamin painted many of the dignitaries in town, art critics have agreed that West did not render two portraits of a man and woman once thought to be Robert's parents. Ironically, the forged West signature appearing on the paintings guarantees their falseness. Following the custom of itinerant portraitists in colonial days, West did not sign his work. Neither the style nor the level of skill match known efforts of his even at this early stage of his development. Furthermore, the subjects are probably not even the Fultons. The clothing style typifies an era several decades later.

13. Letter from London dated October 20, 1805 in the Montagu Collection of the New York Public Library.

CHAPTER 2

1. Andrews may have fled New York as many residents did when the British seized control of the city in September 1776.

2. Flexner, op. cit., pp. 115–116.

3. Most of the tangible monuments to Fulton remain in and near Lancaster, a mere village when he lived there. The city honors his memory by his restored birthplace, the Fulton National Bank, a street name, and his statue in a niche above the Fulton Opera House. In the cities where he would spend most of his life—Philadelphia, London, Paris and New York—a Manhattan street name, a churchyard memorial, and his casket are about all that commemorate him.

4. Flexner, op. cit., p. 117.

5. Fulton probably painted Franklin as many contemporary artists did, but the result has disappeared.

6. Letter from Dr. David Hosack to Cadwallader D. Colden, a close friend of Fulton who invested all his savings in the steamship company. Colden wrote a biography of Fulton in 1817.

7. Fulton repaid it in a typically circuitous way. He, his sisters and brother jointly repaid it with money Robert provided. He then canceled the debt in his will; however, Fulton retained title to the farm until his death.

8. Also spelled Polk or Pollock.

9. Their daughter, Louisa Morris, one of six children, married Dr. Alexander Blair, a surgeon in the United States Army during the War of 1812. Their daughter, Eliza, became the first wife of William Thaw of Pittsburgh.

10. Crumrine, Boyd, *History of Washington County* (Philadelphia, 1882), p. 811 ff.

CHAPTER 3

1. Washington, Pa. and Washington County where Fulton had bought land were named after the general's half-brother who had owned large tracts and had willed them to him upon his death.

2. Sutcliffe, Alice Crary, "Early Life of Robert Fulton," *Century* 76:780–94 (September) 1908.

3. Harlow, Alvin F., *Old Towpaths* (New York: Appleton, 1926), p. 23.

4. Dated September 7, 1784, in Fitzpatrick, John C., editor, *The Writings of George Washington*, Washington, 1931–41, XXVII, p. 468.

5. The Rumsey connection deserves highlighting because the man will figure importantly in Fulton's later life. It is an early example of uncanny coincidences involving Fulton and the steamboat that turn up repeatedly.

6. Flexner, op. cit., p. 120.

7. Boyd, Thomas, *Poor John Fitch* (New York: Putnam's, 1933), pp. 135–136.

8. Fitch published a description of what he intended to do in the *Columbian Magazine* in January 1787. He began this description in the future tense but turned unconsciously to the present as he was so sure that his boat would work as planned.

9. Montagu Collection, New York Public Library.

10. Boyd, op. cit. p. 140.

CHAPTER 4

1. Letter dated April 14, 1789 in possession of Mrs. Frank Semple of Sewickley, Pa.

2. Letter dated May 21, 1793 in the Chicago Historical Society.

3. Letter dated April 14, 1789 in possession of Mrs. Frank Semple of Sewickley, Pa.

4. Letter dated January 20, 1792 in possession of Mrs. Frank Semple of Sewickley, Pa.

5. Ibid.

6. Letter dated July 31, 1789 in possession of Mrs. Frank Semple of Sewickley, Pa.

7. Letter dated January 20, 1792 in possession of Mrs. Frank Semple of Sewickley, Pa. Fulton may have just found rooms at the house of Robert Day, 84 Charlotte Street, Rathbone Place. The Royal Academy of London records show that he also lodged at 67 Margaret Street, Cavendish Square in 1791, and at 18 Newman Street in 1793.

8. Flexner, James T., *America's Old Masters* (New York: The Viking Press, 1939), p. 95.

9. Letter dated January 20, 1792 in possession of Mrs. Frank Semple of Sewickley, Pa.

10. Ibid.

11. Flexner, op. cit., p. 218. The engravings were reproduced in Dickinson, op. cit. pp. 18, 20.

12. Letter dated January 20, 1792 in possession of Mrs. Frank Semple of Sewickley, Pa.

13. Ibid.

14. Gooch, S. W. and Gooch, G. P., *The Life of Charles, 3rd Earl of Stanhope* (London, 1914), p. 168.

CHAPTER 5

1. Dickinson, op. cit., p. 23.

2. Ibid., pp. 24–25.

3. The eliptical spring was not invented until 1804.

4. Letter dated April 1, 1794 in the Chicago Historical Society.

5. This was Fulton's one and only English patent—No. 1988.

6. Dickinson, op. cit., p. 29.

7. Ibid.

8. Ibid., pp. 37–38.

9. In order to announce this forthcoming book, Fulton wrote an article "Small Canals" which was published in the *Star,* July 30, 1795. The dedication of the book was dated March 1, 1796. The publisher was I. and J. Taylor.

10. Letter dated September 12, 1796 in the Historical Society of Pennsylvania, Philadelphia, Pa.

11. Pennsylvania continued to adhere to its plan of turnpike roads,

probably more generally useful than, although as costly as, small canals would have been.

12. Dickinson, op. cit. Letter dated December 28, 1796.

13. Owen, Robert, *Life of Robert Owen* (1857), p. 64.

14. Dickinson, op. cit., p. 35.

15. Ibid., p. 62.

16. From Samuel Coleridge's "The Rhyme of the Ancient Mariner."

17. His baritone voice probably could overcarry them all.

18. Owen, op. cit., p. 70.

19. Memoir of the Rise and Progress of the Chesapeake and Delaware Canal, 1821, Philadelphia, 1821, p. 49.

CHAPTER 6

1. *Memoires de Mme. la Duchesse de Gontaut* (Paris, 1893) pp. 49–50, English translation by Mrs. J. W. Davis (New York, 1894), pp. 63–65.

2. Letter to David Morris dated May 21, 1793 in collection of Chicago Historical Society.

3. *Memoires,* op. cit.

4. Clementia Ross whom Fulton had painted during his apprenticeship days in Philadelphia.

5. Woodress, James, *A Yankee's Odyssey: Life of Joel Barlow* (Philadelphia: J. B. Lippincott, 1958).

6. His epic poem, *Columbiad,* was to be published under the auspices of Fulton.

7. During the Revolutionary War while still an undergraduate at Yale, David Bushnell designed, built and used against the British two weapons that are still mainstays of our nation's armory—the operational submarine and the underwater explosive mine. These technical achievements were apparently conceived *de novo* with no reference to any previous developments.

8. In October 1787 Bushnell sent to Jefferson when he was ambassador to France a detailed description of his experiments, his apparatus and his attempts against the British fleet. The narrative ends in 1778; it may well be the thesis that he submitted to Yale in that year for his M.A. degree.

9. Flexner, op. cit., p. 250.

10. Ibid., p. 247.

11. On December 13, 1797 Fulton submitted definite propositions to the French Director, La Réveillère Lépeaux. This is the Fourth Proposition.

12. From a letter of the Minister of Marine, Georges René Pléville-le-Pelley, who showed the repugnance, which existed universally at that

time, to the employment of what were considered such unfair methods of warfare.

13. Letter to Napoleon dated May 1, 1798, on the eve of his expedition to the East, which ended in the occupation of Egypt. This letter is preserved in the Lenox Library, New York.

14. The end of the report from the commission of experts appointed by the Minister of Marine, Eustace Bruix, to examine Fulton's project dated September 5, 1798.

15. Letter from Fulton to Joshua Gilpin dated Paris, November 20, 1798.

CHAPTER 7

1. The site is now indicated by the "Passage des Panoramas" (namely an arcade with shops).

2. Sutcliffe, op. cit., p. 65.

3. Dickinson, op. cit., p. 100.

4. Ibid., pp. 101–102.

5. Sutcliffe, op. cit., p. 75.

6. Letters from Joel Barlow to Mrs. Barlow, August 17 and 31, 1800.

7. Todd, Charles B., *Life and Letters of Joel Barlow* (New York: G. P. Putnam's Sons, 1886), p. 182.

8. Ibid. Letter from Barlow to Fulton dated September 7, 1800.

9. Capt. S.H. Linzie to Admiralty Secretary, September 21, 1800 in Dickinson, op. cit., p. 109.

10. Report dated November 7, 1800. *Archives Nationales*, Dossier Marine D' 21, fol. 98.

11. Parsons, William B., *Robert Fulton and the Submarine* (New York: Columbia University Press, 1922), p. 37.

12. Fulton to Ministre de la Marine, October 6, 1799, Pesce, G. L. *La Navigation Sous Marine* (Paris, 1906), p. 190. This account includes Fulton's submarine adventures in France.

13. Sutcliffe, op. cit., pp. 80–82.

14. Ibid., pp. 89–95, 320–326.

15. Letter dated September 20, 1801 to the Citizens Monge, La Place and Volney, Members of the National Institute and Commissionaries appointed by the First Consul to promote the Invention of Submarine Navigation.

16. Napoleon made these remarks upon being shown a letter in which Fulton apparently expected the French government to adopt his ideas on his own statement of facts and unverified interpretation of his experiments.

17. Parsons, op. cit., p. 49.

CHAPTER 8

1. A name assumed by Dr. Gregory, emissary from the British government.

2. Flexner, op. cit., p. 295.

3. Entry from Fulton's notebook.

4. Included in the Fulton manuscript left with Consul Lyman to be delivered to Joel Barlow in the event of Fulton's being lost on the voyage home.

5. Parsons, op. cit., pp. 96–97.

6. Ibid., p. 101.

7. Ibid.

8. Cobbett's *Weekly Political Register*, October 27, 1804, p. 640.

9. Memoires, op. cit., pp. 106–107. English translation, op. cit., p. 130.

10. Sutcliffe, op. cit., pp. 168–170. Letter in possession of Judge Peter T. Barlow dated Washington, March 3, 1806.

11. The gallery was established in 1807 and called The Pennsylvania Academy of Fine Arts. *King Lear* eventually found its way to a Boston museum. *Ophelia* is lost.

12. Todd, op. cit., p. 221.

13. Colden, Cadwallader David, *The Life of Robert Fulton* (New York, 1817), p. 217.

14. Dickinson, op. cit., pp. 188–190.

15. Ibid., pp. 193–194.

16. Ibid., p. 196.

17. Parsons, op. cit., p. 125.

18. Ibid., p. 126.

19. Fulton's *Letters Principally to the Rt. Hon. Lord Grenville on Attack*, etc. London, September 23, 1809.

20. Colden, op. cit.

21. Fulton probably encountered stereotyping among printers and engravers while working on the engravings for the *Columbiad*.

CHAPTER 9

1. Spratt, H. Philip, *The Birth of the Steamboat* (Charles Griffin & Co., Ltd., London, 1958), p. 20.

2. Fletcher, R. A., *Steamships* (London: Sidgwick & Jackson, 1910), p. 6.

3. Ibid., pp. 9–10.

4. Spratt, op. cit., p. 25.

5. Rushen, P.C. "History and Antiquities of Chipping Campden in the County of Gloucester," 1899.

6. Flexner, op. cit., p. 18.

7. It is often the neophyte who scores the breakthrough because he doesn't know what can't be done and doesn't have to overcome the barriers of success.

8. Rumsey, James, "A Short Treatise on the Application of Steam," p. 1049. This is from Rumsey's pamphlet attacking Fitch.

9. Flexner, op. cit., pp. 131–132.

10. Fitch did win one influential friend, Dr. William Thornton, a rich patron of science. He would eventually become United States Commissioner of Patents and Fulton's nemisis.

11. Turner, Maria, *James Rumsey, Pioneer in Steam Navigation* (Scottsdale, Pa.: Mennonite Publishing House, 1930), pp. 150–151.

12. Flexner, op. cit., p. 212.

13. Sutcliffe, op. cit., pp. 330–331.

14. Symington, William, "Autobiography," in *Cassiers Magazine* 32:530, 1907.

15. Spratt, op. cit., p. 57.

CHAPTER 10

1. Todd, op. cit., p. 185.

2. $252,000 in terms of today's purchasing power of the American dollar!

3. Turnbull, Archibald D., *John Stevens: An American Record* (New York: Century, 1928), p. 144.

4. Letter from Charles Stoudinger to Nicholas J. Roosevelt, December 29, 1787. New-York Historical Society.

5. Stevens, John, Jr., "A Letter Relative to Steamboats," *American Medical and Philosophical Register* (New York, 1812), 2:419.

6. Flexner, op. cit., p. 285.

7. Todd, op. cit., p. 198. Letter from Barlow to Fulton July 26, 1802.

8. Ibid., pp. 197, 200. Letters dated July 18, August 15, 1802.

9. Miller, Peyton F., *The Story of Robert Fulton* (New York, 1908), p. 96. Fulton manuscript dated February 1813.

10. Robert Fulton to Citizens Molar, Blandell, and Montgolfier, Friends, of the Arts, dated January 25, 1803.

11. Todd, op. cit., p. 189. Barlow to Fulton June 14, 1802.

12. Hill, Ralph N., *Sidewheeler Saga* (New York: Rinehart & Co., 1953), p. 25.

13. Flexner, op. cit., p. 238.

14. Ibid., p. 239.

15. Ibid., p. 240.

16. Fulton to Boulton and Watt, August 6, 1803.

17. Strickland, Mary, *A Memoir of the Life, Writings, and Mechanical In-*

ventions of Edmund Cartwright (London, 1843), p. 150. Fulton to Cartwright March 28, 1802.

18. Letter in possession of C.H. Hart of Philadelphia, Pa.

19. Flexner, op. cit., pp. 291–292.

20. Ibid., p. 293.

CHAPTER 11

1. Flexner, op. cit., p. 317

2. Renwick, James, *Life of Robert Fulton* (New York: Harper, 1845), p. 203.

3. Turnbull, op. cit.

4. S Street between Massachusetts and Florida Avenues now passes through Barlow's former front yard. The house was torn down in 1889.

5. Fulton, Robert, *Torpedo War and Submarine Explosions* (New York: W. Elliott, 1810), p. 7.

6. Dickinson, op. cit., pp. 207–208.

7. Sutcliffe, op. cit., pp. 288–289. Letter in possession of the estate of Cornelia Livingston Crary, Fulton's daughter.

8. Todd, op. cit., p. 233. Fulton to Barlow, spring 1807.

9. Sutcliffe, op. cit., p. 188. Fulton's notebook in possession of Robert Fulton Ludlow.

10. Clarkson, Thomas S., *A Biographical History of Clermont* (Clermont, 1869), p. 135.

11. Also spelled Browne or Brownne.

12. Paulus is also sometimes spelled Pauler's; don't confuse Paulus Hook Ferry with Paulus Hook which is now named Jersey City.

13. In various wordings throughout the Chancellor's correspondence appears the sentiment, "I am sorry that I ever got into this."

14. Dickinson, op. cit., p. 214.

15. Livingston, E. Brockholst and Clermont, *The Livingstons of Callendar* (privately printed, 5:594, 1892).

16. Sutcliffe, op. cit., pp. 202–203.

17. Ibid., p. 212.

18. This is said to have been Fulton's favorite song, and on the one hundredth anniversary of steam navigation, the song was sung upon the decks of the great boats of the Hudson River Day Line.

19. By today's standards, the *North River* handled clumsily.

20. Sutcliffe, op. cit., p. 227.

21. Flexner, op. cit., p. 320.

22. Todd, op. cit., p. 233.

23. Fulton to unknown friend, quoted as part of a reminiscence by the late Judge Story in Sander's *Early History of Schenectady*.

24. Robert Fulton to Robert R. Livingston. Original formerly in possession of Mr. Clermont Livingston.

25. This advertisement costing $4.50 also appeared in the *Evening Post* of New York and included the charges to each passenger and the time to each port; namely, To Poughkeepsie—$4—Time 17 hours.

26. Translated for the *Journal of the Franklin Institute,* Philadelphia, Pa.

27. Account of Judge John Q. Wilson of Albany, N.Y., who in 1856, wrote of his memorable voyage upon the *Clermont,* when she was first put into use as a packet.

28. Sutcliffe, op. cit., p. 274.

29. Letter in possession of a grandson of Captain Andrew Brink.

30. Written in 1857 to J. Franklin Reigart, biographer of Fulton.

31. Ibid.

32. Sutcliffe, op. cit., pp. 213–214.

33. Ibid., pp. 214–215.

34. Thurston, Robert H., *Robert Fulton: His Life and Results* (New York: Dodd, Mead & Co., 1891).

35. Letter of Fulton to Livingston. New-York Historical Society.

CHAPTER 12

1. 1 State Street at the foot of Manhattan Island is now across the street from the dock for the Staten Island Ferry.

2. Woodress, op. cit., p. 254.

3. Ibid., p. 255.

4. Ibid., p. 254.

5. Ibid.

6. From REGULATIONS FOR THE NORTH RIVER STEAMBOAT. Fulton promised to use the fines to buy wine for the company.

7. Trollope, Mrs. Frances, *Domestic Manners of the Americans* (London, 1832).

8. Harlow, op. cit.

9. Dickinson, op. cit., p. 228.

10. Fulton's notes on the Albany Co. Manuscript in Montagu Collection of the New York Public Library.

11. United States of America, the Seventy-Fourth Congress: "Pooling of Patents," *Hearing Before House Committee on Patents on House Resolution 4523,* Parts I–IV, January 3, 1935 to January 3, 1937.

12. Jefferson, Thomas, *Notes on Virginia* in Gilbert Chinard's *The Commonplace Book of Thomas Jefferson* (Baltimore, 1926).

13. Harlow, op. cit., p. 25.

14. Fulton to John De Lacey, March 20, 1814. Manuscript in collection of New-York Historical Society.

15. Sutcliffe, op. cit., p. 269. Letter in possession of C. H. Hart, Esq., of Philadelphia.

16. Turnbull, op. cit.

17. Ibid.

18. Ibid.

19. Ibid.

20. Ibid.

CHAPTER 13

1. Letter of September 15, 1810 to Messrs. Boulton and Watt. Dickinson, op. cit., pp. 230–231.

2. Latrobe, John H.B., *The First Steamboat Voyage on Western Waters* (Baltimore, 1871).

3. Morrison, John H., *History of American Steam Navigation* (New York, 1903), p. 205.

4. Letter in possession of Judge Peter T. Barlow. Fulton asks Barlow to obtain the signature of William Thornton, clerk of the Patent Office, for a deposition.

5. Fletcher, R.A., *Steamships* (London: Sidgwick and Jackson, 1910), pp. 39–40.

6. Svinin, Pavel Petrovich, "Deck Life on the *Paragon*," Metropolitan Museum of Art, New York.

7. Perry, John, *American Ferryboats* (New York: Wilfred Funk, Inc., 1957), pp. 60–61.

8. Letter dated July 12, 1813. Manuscript in collection of New-York Historical Society.

9. Quoted in Harlow, op. cit., p. 28.

10. Sutcliffe, op. cit., pp. 266–267.

CHAPTER 14

1. Turnbull, op. cit., p. 434 ff.

2. New York Public Library Bulletin *13:* 573.

3. Letter to Lord Stanhope on stereotype printing, dated Oct. 4, 1806.

4. From a *Souvenir of the Hudson-Fulton Celebration,* prepared and published by the Free Library of Jersey City.

5. Fulton to Stanhope, April 10, 1811.

6. Dr. Thornton went to a British officer about to give orders to burn the building and told him: "To burn what would be useful to all mankind would be as barbarous as formerly to burn the Alexandrian Library for which the Turks have been ever since condemned by all the enlightened nations." As a result of his persuasiveness, the Patent Office was the only federal building not set to the torch by the British forces.

7. Turnbull, op. cit., p. 44.

8. Fulton to Monroe in *Journal of American History*, 1:427, 1907.

9. Clark, Allen C., "Dr. and Mrs. William Thornton," *Columbia Historical Society Washington Records*, 1915.

10. Fulton to Barlow, June 28, 1811. Letter in possession of Judge Peter T. Barlow.

11. Morrison, John H., *History of American Steam Navigation* (New York, 1903), p. 207.

12. Colden, op. cit., pp. 248–251.

13. Excerpts from all the above letters in Hill, Ralph N., *Sidewheeler Saga* (New York: Rinehart & Co., 1953), pp. 40–42.

CHAPTER 15

1. Fulton, Robert, "Concluding Address of Mr. Fulton's Lecture on the Mechanism, Practice and Effects of Torpedoes," delivered at Washington, Feb. 17, 1810. *American Tracts*, Vol. 110.

2. Dickinson, op. cit., p. 210.

3. Fulton to Stanhope, April 10, 1811.

4. Rowbotham, W.B., "Robert Fulton's Turtle Boat," in Procedures, U.S. Naval Institute, LXII, 1936, p. 1748.

5. Like the monopoly, this financing method would haunt Fulton's heirs for years.

6. Smithsonian Institution Bulletin 240. *Contributions from the Museum of History and Technology*. Paper 39: Fulton's "Steam Battery," 1964, p. 14.

7. *Edinburgh Evening Courant*, Aug. 31, 1815.

8. Flexner, op. cit., pp. 358–359.

9. Smithsonian Institution, op. cit., appendix.

CHAPTER 16

1. Woodress, op. cit., p. 307.

2. Colden, op. cit., p. 265.

3. U.S. Weather Service Records Office, Asheville, N.C.

4. Letter from Georgetown, District of Columbia, Nov. 19, 1829, in possession of the New-York Historical Society.

5. Julia married Charles Blight of Philadelphia and had three children. Mary married Robert Ludlow of Claverack, N.Y., and had one child. Cornelia married Edward Charles Crary of New York City and bore him five children.

6. Old South Leaflets, 5, 1902.

7. Dickinson, op. cit., p. 268.

EPILOGUE

1. Dangerfield, George, "Steamboats' Charter of Freedom: Gibbons vs. Ogden," *American Heritage,* 14:42 (October) 1963.

2. Warren, Charles, *The Supreme Court in United States History* (Boston: Little, Brown, 1926).

Bibliography

Barlow, Joel (the second), private conversations and correspondence with the author.

Beltzhoover, George M., Jr. "James Rumsey, the Inventor of the Steamboat." *West Virginia Historical and Antiquarian Bulletin,* 1900, pp. 16–18.

Boyd, Thomas. *Poor John Fitch.* New York: G.P. Putnam's Sons, 1935.

Buchanan, Lamont. *Ships of Steam.* New York: McGraw-Hill, 1956.

Buckman, David Lear. *Old Steamboat Days on the Hudson River.* New York: Grafton Press, 1907.

Butler, L. L. "He Made the First Steamboats, John Fitch." *Mentor* 16:36–7 (June) 1928.

Caldwell, John Alexander. *Illustrated, Historical Centennial Atlas of Washington County, Pa.* Condit, Ohio, 1876, p. 170.

Carmer, Carl. *The Hudson.* New York: Farrar & Rinehart Inc., 1939.

Carroll, Barbara, B. "Manufacturing in a Young America." *Industry Week,* July 7, 1975, pp. 23–30.

Chapman, William, *Observations on the Various Systems of Canal Navigation.* London, 1797.

Chatterton, E. K. *Steamships and Their Story.* London: Cassell & Co., 1910.

Clark, Allen C. "Dr. and Mrs. William Thornton." *Columbia Historical Society Washington Records, 1915,* pp. 144–208.

Clarkson, Thomas S. *Biographical History of Clermont.* Clermont, 1869.

Cobbett, William. *Political Register,* 4:641–42 (October 27) 1804.

Colden, Cadwallader David. *The Life of Robert Fulton.* New York, 1817.

Creel, George. "Beloved Genius." *Collier's* 77:11–12 (May 1) 1926.

Crumrine, Boyd. *History of Washington County.* Philadelphia: L. H. Evert & Co., 1882.

Dangerfield, George. "Steamboat's Charter of Freedom: Gibbons vs. Ogden." *American Heritage* 14:38–43 + (October) 1963.

Delaplaine, Joseph. Life of Robert Fulton in *Delaplaine's Repository of Lives and Portraits of Distinguished American Characters,* I. Philadelphia, 1816.

Dickinson, Henry W. *Robert Fulton, Engineer and Artist, His Life and Works.* London: John Lane, 1913, reprinted 1971.

Donovan, Frank. *River Boats of America.* New York: Thomas Y. Crowell Co., 1966.

Dorfman, Joseph. "Robert Fulton." *Political Science Quarterly 59* #4, 1929.

Drago, Harry S. *The Steamboaters*. New York: Dodd, Mead, 1967.

Eshleman, H. Frank. "Chronology of Lancaster County." *Lancaster County Historical Society Journal* 27:48, 1923.

Eskew, G.L. "The Father of the Steamboat." *Popular Mechanics* 52:442–7 (September) 1929.

"Fitch vs Fulton for Steamboat Honors." *Literary Digest* 104:36 (March 8) 1930.

Fletcher, R.A. *Steamships*. London: Sidgwick & Jackson, 1910.

Flexner, James T. *Steamboats Come True, American Inventors in Action*. New York: The Viking Press, 1944.

Forrest, Earle. *History of Washington County, Pa.* Vol. I. Chicago: S. J. Clarke Publishing Co., 1926, p. 443.

Fulton, Eleanore J. *An Index to the Will Books and Interstate Records of Lancaster County*. Lancaster, Pa., 1936.

———. "Robert Fulton as an Artist." *Lancaster County Historical Society Journal* 42:49–96, 1938.

Fulton, Harriet Livingston. Letter to Robert R. Livingston, July 29, 1812. New-York Historical Society.

Fulton, Robert. Letter to Sister Elizabeth from London. October 20, 1805. New York Public Library.

———. Letters Principally to the Rt. Hon. Lord Grenville on Submarine Navigation and Attack, etc. London, September 23, 1806, pp. vi, 31–34.

———. "Mechanism, Practice and Effects of Torpedoes." *American Tracts 110*, 1810.

———. *Torpedo War and Submarine Explosions*. New York: W. Elliott, 1810.

———.*Treatise on the Improvement of Canal Navigation*. London, 1796.

Furber, H. "Fulton and Napoleon in 1800: New Light on the Submarine Nautilus." *American Historical Review* 39:489–94 (April) 1934.

Gilpin, Joshua. *Memoir on the Rise and Progress of the Chesapeake and Delaware Canal*. Privately published, 1821.

Gooch, H.W. and Gooch, G.P. *The Life of Charles, 3rd Earl of Stanhope*. London, 1914.

Gontaut-Biron, Marie Josephine Louise de Montaut de Navailles, duchesse de. *Memoires de la Duchesse de Gontaut*. Paris, 1893. English translation by Mrs. J.W. Davis, New York City, 1894.

Harlow, Alvin F. *Old Towpaths*. New York: Appleton, 1926.

Hill, Ralph N. *Sidewheeler Saga*. New York: Rinehart & Co., 1953.

Iles, George. *Inventors at Work*. New York: Doubleday, Page & Co. 1906.

Jefferson, Thomas. *Writings*, vol. v, letter to Benjamin Stiles.

———. *Notes on Virginia*, in Gilbert Chinard's *The Commonplace Book of Thomas Jefferson*. Baltimore, 1926.

Jones, Stacy V. *The Patent Office*. New York: Praeger, 1971.

Lambert, Derek. *Blackstone and the Scourge of Europe.* New York: Stein & Day, 1974.

Lane, Carl O. *American Paddle Steamboats.* New York: Coward-McCann, Inc., 1943.

Latrobe, J. H. B. "A Lost Chapter in the History of the Steamboat." Baltimore: *Maryland Historical Society Fund Publication No. 5,* 1871.

Livingston, Robert R. Letter to editors of *American Medical and Philosophical Register.* Boston: Old South Leaflets No. 108, 1900.

Lowe, Corinne. *Quicksilver Bob.* New York: Harcourt, 1946.

Marine Museum of The City of New York: Collection on Robert Fulton.

Miller, Peyton F. *Story of Robert Fulton.* New York: Knickerbocker Press, 1908.

Mitman, Carl W. "Robert Fulton." *Dictionary of American Biography* 4:68–72, 1932.

Morrison, John H. *History of American Steam Navigation.* New York: 1903.

Nevins, Allan, ed. *Diary of Philip Hone.* New York: Dodd, Mead & Co., 1927.

Owen, Robert. *The Life of Robert Owen.* London, 1857.

Parsons, William B. *Robert Fulton and the Submarine.* New York: Columbia University Press, 1922, reprinted 1967.

Patterson, Andrew, Jr. "David Bushnell, '75: Inventor and engineer of destruction." *Yale Alumni Magazine* 39:36–7 (October) 1975.

Peale, Charles Willson. Correspondence with Robert Fulton. *Pennsylvania Magazine of History and Biography* 9:129, 1889.

Perry, John. *American Ferryboats.* New York: Wilfred Funk Inc., 1957.

Pesce, G.L. *La Navigation Sous Marine,* Paris, 1906.

Podmore, Frank. *Robert Owen, A Biography.* London: George Allen & Unwin Ltd., 1906, reprinted 1968.

Presbytery of Washington County. *The Cross Creek Presbyterian Church,* 1889, pp. 5–11.

Reigart, J. Franklin. *The Life of Robert Fulton.* Philadelphia: C. G. Henderson & Co., 1856.

Reninger, Marion Wallace. "The Great Ideas of Robert Fulton." *Lancaster County Historical Society Journal* 72:170–185, 1968.

Renwick, James. *Life of Robert Fulton.* New York: Harper, 1845.

Rips, Rae Elizabeth. *U.S. Government Manual,* 74–75. New York: H. W. Wilson Co., 1949.

Roosevelt, Hall and McCoy, S.D. *Roosevelt: Odyssey of an American Family.* New York: Harper, 1939.

Rumsey, James. "A Short Treatise on the Application of Steam," reprinted in Documentary History of the State of New York. Albany, 1849.

Rushen, P. C. "History and Antiquities of Chipping Campden in the County of Gloucester," 1899.

Smith, Sidney. Letter to Viscount Castlereagh, November 22, 1805; in Castlereagh *Correspondence,* London 5:131, 1851.

Smithsonian Institution Bulletin 240. *Contributions from the Museum of History and Technology.* Paper 39: Fulton's "Steam Battery," 1964.

———. Washington, D.C.: *Annual Report,* 1929.

Sparks, Jared. *Gouverneur Morris.* Boston: Gray & Boeden, 1832.

Spratt, H. Philip. *The Birth of the Steamboat.* London: Charles Griffin & Co., 1908.

Strickland, Mary Cartwright. *A Memoir of the Life, Writings and Mechanical Inventions of Edmund Cartwright.* London, 1843.

Sueter, Commander Murray F. *Evolution of the Submarine Boat.* London: J. Griffin & Co., 1908.

Sutcliffe, Alice Crary. "Early life of Robert Fulton." *Century* 76:780–94 (September) 1908.

———. *Robert Fulton.* New York: Macmillan, 1915.

———. *Robert Fulton and the* Clermont. New York: Century, 1909.

Symington, William. "Autobiography," in *Cassiers Magazine* 32:530, 1907.

Thurston, Robert H. *Robert Fulton: His Life and Results.* New York: Dodd, Mead & Co., 1891.

Todd, Charles B. *Life and Letters of Joel Barlow.* New York: G. P. Putnam's Sons, 1886.

Turnbull, Archibald D. *John Stevens. An American Record.* New York: Century, 1928.

Turner, Maria. *James Rumsey, Pioneer in Steam Navigation.* Scottsdale, Pa.: Mennonite Publishing House, 1930.

United States of America, The Seventy-Fourth Congress: "Pooling of Patents," *Hearings before House Committee on Patents on House Resolution 4523,* Parts I–IV. January 3, 1935 to January 3, 1937.

———. The Seventy-Fifth Congress: "Compulsory Licensing of Patents." *Hearings before House Committee on Patents on House Resolutions 9259, 9815, 1666.* January 3, 1937 to January 3, 1939.

United States Weather Service Records Office, Asheville, N.C.

Vaughan, Floyd L. *The United States Patent System.* Norman, Okla.: Oklahoma Press, 1956.

Warren, Charles. *The Supreme Court in United States History.* Boston: Little, Brown, 1926.

Whiteman, Maxwell. *Copper for America.* New Brunswick, N.J.: Rutgers University Press, 1971.

Woodress, James. *A Yankee's Odyssey: Life of Joel Barlow.* Philadelphia: J. B. Lippincott, 1958.

Wynne, James. *Lives of Eminent Literary and Scientific Men of America.* New York: Appleton, 1850.

INDEX